The Pea
& the Sun

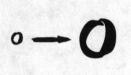

The Pea & the Sun

A Mathematical Paradox

Leonard M. Wapner

CRC Press
Taylor & Francis Group
Boca Raton London New York

CRC Press is an imprint of the
Taylor & Francis Group, an **informa** business

AN A K PETERS BOOK

Editorial, Sales, and Customer Service Office

First published 2005 by A K Peter, Ltd.

Published 2023 by CRC Press
Taylor & Francis Group
6000 Broken Sound Parkway NW, Suite 300
Boca Raton, FL 33487-2742

ISBN-13: 978-1-56881-327-1 (pbk)

Library of Congress Cataloging-in-Publication Data

Wapner, Leonard M., 1948-
  The pea and the sun : a mathematical paradox / Leonard M. Wapner.
     p. cm.
  Includes bibliographical references and index.
  ISBN 1-56881-213-2
    1. Banach-Tarski paradox--History. I. Title.

QA248.W29 2005
511.3'22--dc22

2004063620

Visit the Taylor & Francis Web site
at http://www.taylorandfrancis.com .

and the CRC Press Web site at
http://www.crcpress.com

To Kirsty
with love,
Dad

Table of Contents

Acknowledgments

Family first! I thank my mom, Lea, and my dad, Ben, for *starting this project*. Thanks Mom, for the love, support, and encouragement to write. Now you write *your book*! Thanks Dad, for your love and for teaching me so many things. I miss you. My wife, Mona, deserves a book of thanks for her love and support. Maybe I'll write one; but for now and always—I love you! And for our daughter Kirsty, thanks for always *being you*. I love you all very much.

I am indebted to many and it is with great pleasure that I acknowledge those who have helped me write this book. I begin by thanking my teachers from elementary school through graduate study. I have modeled my teaching style after many of these teachers and I hope to inspire my students as my teachers have inspired me.

Since elementary school I've enjoyed reading popular (general audience) books about science and mathematics. These books and articles played no small part in my deciding to study and ultimately teach mathematics. Of all that I have read and continue to read, I single out Martin Gardner as having the most profound influence in my choice of study and profession. The reader may know that Mr. Gardner served as the Mathematical Games columnist for *Scientific American* from 1956 to 1981 and has written over 70 books on various topics. Though having no formal academic status in mathematics (his degree is in philosophy), he has received praise from mathematicians worldwide and I'm certain that I'm not alone in owing him a great debt of gratitude.

Acknowledgments

I thank my colleagues at El Camino College for patiently answering my questions and offering assistance. (They had no choice as I can be persistent.) I am particularly grateful to Bill Hemmer for encouraging me to continue writing at a time when I was considering giving up on this project. Bob Lewis deserves credit for convincing me the Cantor set is indeed remarkable. I thank my good friend and colleague Paul Wozniak for reviewing parts of the manuscript and offering constructive criticism. And I owe much to Michael Berg, professor of mathematics at Loyola Marymount University, undoubtedly one of the most talented mathematicians I know.

With the power of email I have apologetically imposed myself upon some world class authorities on this subject matter. Jan Mycielski, Stan Wagon, Matthew Foreman, Robert French, Karl Svozil, and Yiannis Moschovakis have graciously answered my questions and most have been cited in this book. I thank them all.

In this digital age where information is no more than a mouse click away via the Internet, there remains no more reliable a source than the *bricks and mortar* library. I have visited many while working on this book and I am especially indebted to the administration, reference librarians, and support staff of the Science Library at the University of California, Irvine. They are a talented and patient group whose assistance has been invaluable.

I give special thanks to my longtime friend, Howard Glassman, for answering an endless stream of computer questions and helping with the graphics.

Finally, I wish to thank Alice and Klaus Peters, along with the staff at A K Peters, Ltd., for their trust, expertise, and guidance throughout this project. I offer a special thanks to Charlotte Henderson for editorial assistance, Darren Wotherspoon for design and production, and Susannah Sieper for marketing.

Introduction

Dear God—
If I have but one hour remaining to live,
please allow me to spend this time in a mathematics class
so that it will seem to last forever.

—A bored student's prayer

The clock had stopped and I was falling asleep.

In 1971, I was completing my graduate study in mathematics at UCLA, taking a class entitled *Measure and Integration.* I was bored. Despite the fact my major was mathematics, this particular course held no interest for me. I had then, and retain today, a strong love of mathematics and have spent most of my adult life teaching the subject at El Camino College. This course was a requirement for my major, however, and I just wanted it out of the way. I had no idea as I stared at the clock that I was about to be introduced to a truly remarkable theorem.

The professor was concluding his lecture and summed up by drawing a solid sphere (ball) on the board. He claimed he had just presented a proof of the fact that the ball could be partitioned into five pieces and then rearranged, much like a jigsaw puzzle, in such a way that two balls could be formed, each identical in shape and volume to the original.

I was immediately reminded of those red sponge balls a magician uses. He places one ball in his hand and closes it. When he opens his hand, there are two balls. Or doves? He puts one dove in the box,

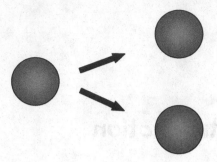

Figure I.1. Two solid spheres from one.

closes the lid, opens it and out fly two doves! So, I assumed there was some sort of joke or trick to this; I hadn't been paying attention and I had no idea as to the professor's intentions.

He continued. "So, we have the Banach-Tarski Theorem, or Banach-Tarski Paradox. An equivalent form of this theorem states that a solid of any size, say that of a small pea, can be partitioned into a finite number of pieces and then reassembled to form another solid of any specified shape and volume, say that of the sun. Consequently, this paradoxical theorem of Stefan Banach and Alfred Tarski is sometimes referred to as the *pea and the sun* paradox." (See Figure I.2.)

Is it conceivable *aus einer Mücke einen Elefanten machen*? Can an elephant be made out of a mosquito? It was not April Fools' Day! Was this a joke? And if so, what was the punch line? I was unwilling to hazard a potentially embarrassing question so I just looked around the room to see how others were taking all of this.

The student on my right raised his hand. "Clearly these results are nonsense, right? I mean, you're not suggesting that one can cut an apple into five pieces and reassemble the pieces to form two apples, are you? Are you suggesting that we can create something out of nothing?"

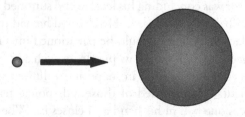

Figure I.2. The pea and the sun.

By now I had stopped watching the clock and was staring intently at the professor. I expected he would deliver a humorous punch line and dismiss the class or explain why, in fact, the theorem's proof was flawed. Instead, he just dismissed the student's questions with a blasé response.

"Well, you know, it's just one of those things. Results like this are common when we work with nonmeasurable sets, the Axiom of Choice, etc. The proof is valid and the theorem is accepted."

The class ended, and I left feeling a bit confused.

This was my introduction to the remarkable Banach-Tarski Theorem. In the foreword to *The Banach-Tarski Paradox* by Stan Wagon, Jan Mycielski refers to this theorem when he writes [Wagon 85, p. xi], "This, I believe, is the most surprising result of theoretical mathematics." For obvious reasons, the theorem's publication in 1924 was followed by a storm of controversy among mathematicians. How could such results, blatantly contradicting common sense, be accepted?

As the general population learned of the theorem, the controversy spread. An irate citizen once demanded of the Illinois legislature that they outlaw the teaching of this result in Illinois schools [Addison 83, p. 28]. So, two camps evolved—one accepting the beautifully counterintuitive results and the other rejecting it all as being meaningless.

Despite the proliferation of highly technical journal articles on the subject, little has been written for the general public. Indeed, browsing the library stacks and searching the Internet yield little of interest to anyone not having a graduate degree in mathematics. As a result, there is significant misunderstanding as to the nature of the theorem. I've had students ask, "What's this I hear about mathematicians being able to duplicate matter? They proved some sort of theorem suggesting we can build duplicating machines, right?"

The spirit of this book is to make this topic accessible by providing a journalistic, as opposed to mathematically intensive, look at the theorem. Chapter 1 addresses historical matters by presenting the *cast of characters*. The stars are Georg Cantor, the founder of modern set theory; Stefan Banach and Alfred Tarski, the leads; Kurt Gödel, the foremost logician of the twentieth century; and finally, Paul Cohen, professor of mathematics at Stanford University, who puts a sense of closure on the matter.

Chapter 2 presents a collection of mathematical recreations involving the geometrical dissection and reassembly of figures where something is magically gained, or lost, in the process. Though clever in construction, it must be stressed that these are included as recreations in comparison to the mathematically sound Banach-Tarski Theorem.

Chapter 3 presents the prerequisite mathematics necessary to fully appreciate the theorem. Being written for a general audience having a mathematics background including algebra and geometry, mathematical formality is excluded. Readers interested in pursuing the subject and having the appropriate mathematics background will find the bibliography helpful.

I whimsically entitle Chapter 4 "Baby BTs." These are mathematical curiosities, involving an apparent gain by decomposition and reassembly. Mathematically, they are somewhere between the jigsaw recreations given in Chapter 2 and the Banach-Tarski Theorem.

In Chapter 5, the statement and proof of the Banach-Tarski Theorem are given. Mathematical formality is omitted to reach a wider audience. This is not done at the expense of correctness.

The resolution of the paradox is given in Chapter 6. In some ways, a paradox loses its character, once resolved; but no discussion of this beautiful theorem would be complete without some explanation of its magic. Magicians and mathematicians treat similar mysteries in opposite ways. Magicians never reveal secrets; mathematicians strive to expose and clarify secrets.

Is there a physical reality to the consequences of the Banach-Tarski Theorem, or has mathematics just gone off the deep end? Chapter 7 will provide some answers.

Chapter 8 closes the presentation with a look at the past and future of mathematical discovery.

As a lifelong mathematics educator, I have more respect for questions than answers. Therefore, if the reader is to conclude this book with more questions than when he or she began, I will be gratified.

Len Wapner

1 History: A Cast of Characters

The good Christian should beware of mathematicians,
and all those who make empty prophecies.
The danger already exists that mathematicians have
made a covenant with the devil to darken the spirit
and to confine man in the bonds of Hell.

—St. Augustine

Significant mathematical achievement is best understood when viewed in correct historical and mathematical context. Mathematics is not created in a vacuum and, in the case of Banach and Tarski's remarkable work, there are at least three other mathematicians to acknowledge—Georg Cantor, Kurt Gödel, and Paul Cohen. This chapter gives a four part history of the Banach-Tarski Theorem:

1. Georg Cantor introduces the concepts of set theory and transfinite arithmetic.

2. Stefan Banach and Alfred Tarski publish the Banach-Tarski Theorem.

3. Kurt Gödel shows the Axiom of Choice is consistent with the other axioms set theory.

4. Paul Cohen shows the Axiom of Choice is independent of the other axioms of set theory.

With a nod to Jules Verne, this chapter could have been entitled "Around the World in Eighty *Years*," as it is truly an international story with Germany, Poland, Austria, and the United States well represented.

Before beginning, let's briefly consider the nature of mathematical achievement. Do mathematicians discover or create? Did Stefan Banach and Alfred Tarski discover their mysterious paradox lurking in the depths of mathematical truth, or did they create it? Similar questions could be asked of the scenic or portrait photographer. Environmentalist Ansel Adams is best known for his stunning black and white photographs of Yosemite Valley and the Sierra Nevada mountains in California. Did he *record* or *create*? As black and white images with extreme light and shadow effects, the photographs are not realistic impressions of what one actually sees when hiking these mountains. In this sense, the photographs are created (artistically) with a good eye and state of the art equipment. But these beautiful photographs may also represent nature at its best and one gets a sense of seeing *God's gifts* when looking at Adams' photographs. Did Adams create this beauty, or simply discover and record it with the right technique and equipment?

With respect to mathematics, there are two opposing viewpoints. The Platonic view of mathematics—*Platonism* (or *mathematical realism*)—holds that mathematical objects exist *out there*, independent of the human mind. "Pi" in the sky! Theorems, proofs, constructions, and solutions to unsolved problems are waiting to be discovered by mathematical researchers much the same way gemstones are waiting to be unearthed by geologists. According to the Platonist, mathematical shapes, quantities, and relationships have always existed, at least in a theoretical sense. They are no more creations of the human mind than a diamond is a creation of the geologist. A talented mathematician does not create mathematics. Mathematics is discovered. Thus, it may be that the popular view of the mathematician is that of a Platonist, as many think of the mathematician as a scientist, rather than a creator or artist. Georg Cantor and Kurt Gödel have generally been regarded as Platonists. It has been suggested that most of today's mathematicians are Platonists, but few are willing to admit it. Mathematician and human rights advocate Lipman Bers has stated [Albers, Alexanderson, and Reid 90, p. xiii], "A working mathematician is always a Platonist. It doesn't matter what he says . . . I think that in mathematics he always has that feeling of discovery . . . Mathematics is, as Ron Graham has said, the ultimate reality . . . ."

The opposing viewpoint, known as *formalism*, holds that mathematics is a language consisting of symbols, and conventions for manipulating these symbols which, when the *rules of the game* are followed, generate theorems, proofs, constructions, etc. It is a construct of the human mind. These theorems, sets, etc. need not be applied to the physical world. Mathematics stands separated from physical reality as a human creation, much like a spoken language or work of art.

A third viewpoint is that of the *constructivist* (*intuitionist, finitist*), believing that only mathematical objects which can be constructed in a finite way have meaning. Constructivists tend to oppose infinite processes and existence theorems which do not construct the object being considered. Think of Platonists and formalists as having opposing viewpoints (see Figure 1.1), with constructivists having issues with both groups.

Are you, the reader, a Platonist? Read the following and decide for yourself.

The number π is the ratio of the circumference of a circle to its diameter. It has been proven that π is irrational. (A real number is rational if it is the quotient of two integers. If a number is rational, it will have a repeating or terminating decimal expansion. If a real number is not rational, it is called irrational.) In fact, it has been proven that π is transcendental, meaning that it is not the root of any algebraic equation. Consequently, it is impossible to compute π (the decimal expansion π) exactly by algebraic means (addition, subtraction, multiplication, division, square roots, etc.). Because there are ways to represent π as the sum of an infinite series (e.g. $\pi = \frac{4}{1} - \frac{4}{3} + \frac{4}{5} - \frac{4}{7} + \ldots$) it is theoretically possible to calculate π to any specified number of decimal places. Despite the fact computers have calculated π to trillions of decimal places, there remain many open questions regarding possible patterns in the decimal expansion of π. Do all digits occur infinitely often? Do they occur with equal frequency? Are there patterns of digits within the expansion of π? There are endless questions of this sort, some of which may never be answered.

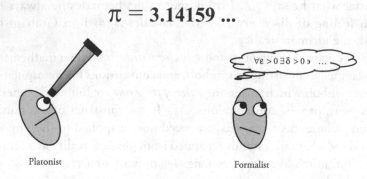

Figure 1.1. "Pi" in the sky.

Later in this chapter when considering the work of Kurt Gödel, we will see that there are mathematical questions and conjectures which are independent of our axiomatic system. That is, there are questions which can never be answered, and are called *undecidable*. For the sake of this test, let's assume that the following question has been proven undecidable: "In our axiomatic system, does the decimal expansion of π contain infinitely many zeros?"

Do you, the reader, believe the decimal expansion of π contains infinitely many zeros?

If your answer is either yes or no, then you are a Platonist. The Platonist sees the decimal expansion of π as being out there, somewhere, and acknowledges the fact that our axiomatic system will never yield an answer to our question. However, the answer must exist, and there may be additional mathematical evidence, perhaps in the form of an extended or alternative axiomatic system, which will decide the issue.

The formalist and constructivist dismiss the question as meaningless. If we do not have the means to obtain the answer, then there is no answer. (Philosophically, this is reminiscent of the old question, "If a tree falls in the forest and there is no one there to hear it fall, does it make a sound?")

So decide for yourself. Are you a Platonist?

Bear in mind that Platonists and formalists work on similar mathematical problems and tend to agree on most mathematical matters. Both prove new theorems, find new proofs to old theorems, and find solutions to unsolved problems. Each group generally accepts the mathematics of the other, but they disagree as to what it all ultimately represents.

Figure 1.2. Georg Cantor (1845–1918).
(From UA Halle Rep. 40 I C 11. Reprinted with the permission of
Martin-Luther-Universität Halle-Wittenberg.)

The five major players of the Banach-Tarski story are (chronologically) Georg Cantor, Stefan Banach, Alfred Tarski, Kurt Gödel, and Paul Cohen. The story is international, beginning in Germany, moving on to Poland, then to Austria, and concluding in the United States. We begin with the story of Georg Cantor, a Platonist by his own admission, and generally recognized as the founder of modern set theory.

Georg Cantor—The Founder of Modern Set Theory

Georg Cantor (see Figure 1.2), born in St. Petersburg in 1845, revolutionized mathematics in establishing set theory (*Mengenlehre*) as a mathematical discipline. In doing so, he was able to actualize or consummate the notion of infinity creating an arithmetic of infinities— transfinite arithmetic. As a Platonist with theological interests, he often saw himself as a secretary or messenger for God.

The proof of the Banach-Tarski Theorem, as presented in Chapter 5, requires the manipulation of infinitely many points of the solid sphere and the manipulation of infinitely many rotations of these points. Without Cantor's revolutionary ideas of set theory and transfinite arithmetic, Stefan Banach and Alfred Tarski would surely not have published their theorem. In fact, twentieth century mathematics would not exist, as we know it.

Dreams of infinite space and time come naturally to anyone gazing at the heavens or contemplating the periodic cycles of day into night. Infinity is historically mystical, having been contemplated by scientists,

5

philosophers, and theologians, as well as mathematicians. By its very nature, it is difficult to define, and in some discussions it may be convenient to reject the concept as meaningless. The history of infinity may be infinite in itself. Cantor was certainly not the first to consider the concept; so, to fully comprehend what Cantor accomplished, we begin with the fourth century BC Greek philosopher, Zeno.

Known for his paradoxes of motion and continuity, Zeno was one of the first to pose serious questions about infinite processes. His Dichotomy Paradox asserts that a runner can never reach the end of his race; for in order to do so he must first reach the halfway mark, then the halfway mark of the remaining half, and so on. So the fraction of the course completed can be thought of as the infinite sum $1/2 + 1/4 + 1/8 + \ldots$. Since infinitely many fractional parts of the course would have to be completed in a finite amount of time, the end of the course could never be reached. Surely it is impossible to complete an infinite amount of tasks in finite time. Right? Another well known paradox of Zeno involves Achilles racing a tortoise with the tortoise being given a head start. Zeno argues that Achilles can never overtake the tortoise. To do so, Achilles must first reach the point where the tortoise started, by which time the tortoise will have advanced to another point. When Achilles reaches that point, the tortoise will have advanced again to a third point. Since the process goes on ad infinitum, there is no hope that the tortoise can be overtaken in a finite amount of time.

Clearly Zeno knew that in actuality the runner of the Dichotomy Paradox completes the course and that Achilles would catch up to the tortoise. Yet, he made no attempt to resolve these paradoxes. It would be over two thousand years before mathematics would resolve such problems.

It was Aristotle in the third century BC who first made the distinction between the actual infinite and the potential infinite. Aristotle writes [Aristotle, 207b]:

> Hence this infinite is potential, . . . and not a permanent actuality but consists in a process of coming to be, like time . . . . With magnitudes the contrary holds . . . In point of fact they (mathematicians) do not need the infinite and do not use it. They postulate only that the finite straight line may be produced as far as they wish . . . Hence, for the purposes of proof, it will make no difference to them to have such an infinite instead, while its existence will be in the sphere of real magnitude.

Thirteenth century Christian theologian Saint Thomas Aquinas writes [Aquinas, Ia 7.4.]:

> The existence of an actual infinite multitude is impossible. For any set of things one considers must be a specific set. And sets of things are specified by the number of things in them. Now no number is infinite, for number results from counting through a set of units. So no set of things can actually be inherently unlimited, nor can it happen to be unlimited.

The great German mathematician Carl Friedrich Gauss (1777–1855), considered by some as the greatest mathematician the world has known, wrote in a letter to a friend [Maor 87, p. 55]:

> I must protest most vehemently against your use of the infinite as something consummated, as this is never permitted in mathematics. The infinite is but a *façon de parler*...

And to this day, mathematics education through geometry, algebra, and calculus generally treats the infinite not as a real number, but rather as the quality of being unlimited in size. Elementary mathematics almost never makes such statements as "$x = \infty$." In its place we see "$x \to \infty$," suggesting infinity as a potential, rather than achievable quantity.

(The popular symbol ∞ used for infinity was first introduced by the English mathematician John Wallis in the seventeenth century. He may have taken it from the Roman numeral for 100 million, which consists of the lazy eight placed within a rectangle. The symbol itself is symbolic of an endless process, perhaps that of a snake devouring itself.)

While mathematicians and philosophers before Georg Cantor could only look to infinity with mathematical telescopes, treating it as a potential, Cantor consummated, or actualized the infinite, dropping it in our laps to be manipulated and explored. His work, revolutionary in its time, brought with it controversy and sadness which ultimately may have contributed to his death.

Born to parents of Jewish descent, Georg moved with his family to Frankfurt, Germany in 1856. His father had converted to Protestantism and his mother was born Catholic. His father encouraged him to pursue a career in engineering but the young Cantor was more interested in the philosophy of medieval theologians, showing interest and talent in philosophy, physics, and mathematics. In the end, his father gave him permission to pursue a career in mathematics. Cantor was grateful and would always feel a need to live up to his father's expectations. Some have suggested that this pressure may have contributed to Cantor's mental health problems later in life.

He earned his doctorate at the University of Berlin in 1867 with a thesis in number theory, then took an entry level position as *Privatdozent* at the University of Halle. It was a low level position at an institution lacking reputation. He was promoted to associate professor and then to professor of mathematics, yet never achieved his dream of a professorship at the University of Berlin. For this failure he blamed his lifelong archenemy Leopold Kronecker (1823–1891), who was highly critical of Cantor's concept of the infinite.

The concept of set, as developed by Cantor, became entwined with the infinite when he began to consider the size, or *cardinality*, of sets. The cardinality of a finite set simply is the number of elements in the set. So, the cardinality of {2, 4, 6} is 3. But what could be said of the cardinality of the set of counting numbers {1, 2, 3, . . .} or of the set of all real numbers? Do they have infinite cardinalities? Are their cardinalities equal? Does it make sense to say that some infinities are greater than other infinities?

To answer such questions, Cantor looked first to finite sets and pointed out that two finite sets would have the same cardinality if their members could be put in one-to-one correspondence with each other. So, the sets {2, 4, 6} and {8, 9, 10} have the same cardinality (3) because of the correspondence $2 \leftrightarrow 8, 4 \leftrightarrow 9, 6 \leftrightarrow 10$. He then suggested the same could be said of infinite sets. For example, the set of even numbers {2, 4, 6, . . .} would have the same infinite cardinality as the set of counting numbers {1, 2, 3, . . .} because there is a clear one-to-one correspondence between the two sets (Figure 1.3).

So, despite the fact that the even numbers form a proper subset of the counting numbers, both sets contain the same (infinite) number of elements and are said to be of the same cardinality. (A proper subset of a given set is a subset of the given set not equal to the given set.) This would be the first of many set theoretic paradoxes as it seemingly contradicts Euclid's common notion of the whole being greater than the part. Paradoxical as this may be, Cantor used this as the very definition of an infinite set. He defined an infinite set as one which could be put in one-to-one correspondence with a proper subset of itself, removing all vagueness from previous notions of infinitely large sets.

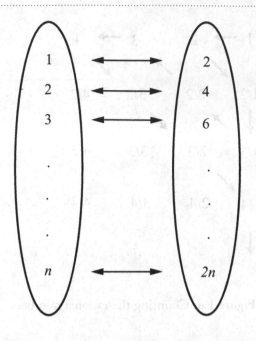

Counting Numbers   Even Numbers

Figure 1.3. One-to-one correspondence between two infinite sets.

Other sets having the same cardinality as the counting numbers include the set of odd numbers, the set of all perfect squares, and the set of primes. In each case, the members of the set are *denumerable*, or *countable*, in that they can be *listed*, putting them in one-to-one correspondence with the counting numbers. The question naturally arises if all infinite sets have the cardinality of the counting numbers: is there a one-to-one correspondence between the rational numbers and the counting numbers? What about the set of all real numbers? Cantor hypothesized a ranking, or hierarchy of infinities, and thus was born the subject of *transfinite arithmetic*, the arithmetic of the infinities.

To investigate, Cantor considered the set of positive rational numbers—all positive numbers which can be written as a ratio of two integers. Cantor had reason to believe that this set might be more numerous than the set of counting numbers, because between any two rational numbers there exists another rational number. In fact, between

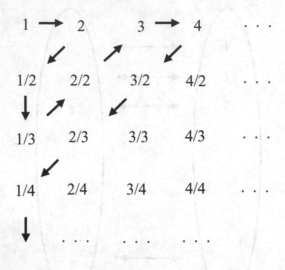

Figure 1.4. Counting the rational numbers.

any two rationals there exists infinitely many rationals. Mathematicians use the word *dense* to describe such sets. In contrast, such sets as the counting numbers, even numbers, perfect squares, and prime numbers all have gaps between their members. (Counting numbers are one unit apart, even numbers are two units apart, etc.)

Cantor was stunned himself to discover that the set of rational numbers can be placed in one-to-one correspondence with the counting numbers, by a diagonal weaving method (see Figure 1.4).

The top row of this array shows all rational numbers, in order of magnitude, with a denominator of one. (We omit the denominator for the first row.) The second row shows all rational numbers, in order of magnitude, with a denominator of two. The pattern continues, horizontally and vertically, in such a way that all positive rational numbers have their position in the table. By following the arrows, we could list these rational numbers in order of succession, taking care to omit any number which may have appeared previously in another form. (For example, we begin by listing "1" and then omitting future repetitions of this value, occurring in the form 2/2, 3/3, etc.) So, our list would appear as

1, 2, 1/2, 1/3, 3, 4, 3/2, 2/3, 1/4, . . .

Clearly every positive rational number would appear once and only once in this sequence. Referring to this sequence, there would be a first rational number (1), a second rational number (2), a third rational number (1/2), and so on. The set of positive rationals is now countable as they have been put in one-to-one correspondence with the set of counting numbers. So, despite the fact that the set of rationals is dense, there are as many rational numbers as there are counting numbers. Cantor found his own results difficult to comprehend, but, as a true Platonist, felt compelled to accept one strange conclusion after another. He pleaded with others not to "blame the messenger."

If a dense set, such as the rational numbers, has the same cardinality as that of the counting numbers, then perhaps all infinite sets are of this same cardinality and there would be, in this sense, only one infinity. Cantor showed otherwise.

He began by restricting the investigation to the real numbers between zero and one. Geometrically these numbers would be represented by points on a line segment of length one. Note that every such number can be represented by a non-terminating decimal fraction. (In the event a decimal representation does terminate, it could be replaced by an equivalent non-terminating representation. For example, the number 3/10, commonly represented as .3, is equivalently represented as .2999999 . . . .) Cantor showed that all real numbers between zero and one could not be put into a one-to-one correspondence with the counting numbers by showing that the existence of such a correspondence would lead to a contradiction.

The proof begins by assuming the correspondence exists. Assume the list, or sequence, is as shown in Figure 1.5.

$$\#1) \quad .a_1 a_2 a_3 \ldots$$

$$\#2) \quad .b_1 b_2 b_3 \ldots$$

$$\#3) \quad .c_1 c_2 c_3 \ldots$$

$$\downarrow$$

Figure 1.5. Cantor's diagonalization proof.

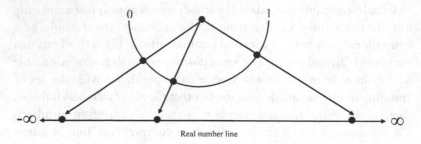

Figure 1.6. Cardinality of the continuum.

Now Cantor constructed a real number $x = .x_1 x_2 x_3 \dots$ in such a way that $x_1 \neq a_1, x_2 \neq b_2, x_3 \neq c_3, \dots$. That is, a real number is created by systematically changing each of the digits along the main diagonal of the array and forming the number x. The newly created number x differs by at least one decimal digit from every number on the list. It follows that x itself can not be on the list, suggesting that the list is incomplete. This contradicts the assumption by which the list was created. So, the set of real numbers between zero and one is not countable. Cantor denoted the cardinality of this infinite continuum by the letter c.

It is relatively simple to show how c also represents the cardinality of all real numbers.

This is done by taking a line segment of length one (consisting of a set of points of cardinality c) and bending it into a semicircle. Figure 1.6 shows a one-to-one correspondence between the set of numbers between zero and one and the entire real number line.

So now Cantor had established the existence of at least two infinities: the first being the cardinality of the counting numbers and the second being the cardinality of the continuum, whether it be all numbers between zero and one or the entire real number line. Could there be even more infinities, some being larger than others? Cantor hypothesized a hierarchy of infinities (thus the name transfinite) denoted by subscripting the first letter of the Hebrew alphabet— $\aleph$ —*aleph*. The smallest infinity, the cardinality of the counting numbers, would be $\aleph_0$ and greater infinities, in sequence would be $\aleph_1$, $\aleph_2$, etc. So he theorized $\aleph_0 < \aleph_1 < \aleph_2 < \dots$ etc. In Chapter 3, we will see at least one way of constructing such a hierarchy. Specifically, we will see that c, the cardinality of the continuum, can be thought of as $2^{\aleph_0}$ (the product formed by using 2 as a factor a countable number of times).

Cantor wondered if there could be an infinite cardinality between $\aleph_0$ and c. He believed (but could not prove) the answer to be "no" thus suggesting the equations $c = \aleph_1$ or $2^{\aleph_0} = \aleph_1$. This equation, or conjecture, known as *Cantor's Continuum Hypothesis*, was elevated to high status in 1900 by the German mathematician David Hilbert (1862–1943) when he placed the question of the Continuum Hypothesis as the number one problem on his famous list of twenty-three unsolved problems. The list was presented by Hilbert at the Second International Congress of Mathematicians held in Paris. In Hilbert's view, the problems encompassed the issues which needed to be addressed by twentieth century mathematicians. To date, more than half of these problems have been solved. To solve one of these problems would be a great honor for any mathematician. Though obsessed with proving (or disproving) the conjecture, Cantor was unable to resolve matters one way or the other. As we see at the end of this chapter, it was not until 1963 that Paul Cohen of Stanford University resolved the matter in a most surprising way.

In actuality, the first problem of Hilbert's famous list consists of two related parts. The first part deals with Cantor's Continuum Hypothesis, as described above. The second part asks if the set of real numbers can be reordered so as to be considered a *well-ordered set*. (A well-ordered set is one in which every nonempty subset has a first element. The conventional ordering of the set of real numbers is not a well ordering in that a subset such as all numbers greater than zero does not have a first element.) It can be shown that the Well-Ordering Theorem, which states that every set can be well-ordered, is equivalent to the Axiom of Choice, a critical component to the proof of the Banach-Tarski Theorem. In other words, contributing to, or solving Hilbert's first problem would have direct impact, one way or the other, on how one should interpret Banach and Tarski's result.

With revolution comes conflict and the revolutionary concepts of Cantor's set theory and transfinite arithmetic began to yield apparent contradictions and controversy. (Specific set theoretic paradoxes as noted by Cesare Burali-Forti, Bertrand Russell, and Cantor himself will be given in Chapters 2 and 3.) German mathematician Felix Hausdorff (1868–1942) describes [Hausdorff 14, p. v] the subject as "a field in which nothing is self evident, whose true statements are often paradoxical, and whose plausible ones are false." The French mathematician Henri Poincaré (1854–1912) referred to Cantor's set theory as "a malady, a perverse illness from which some day mathematics would be cured" [Aczel 00, p. 111]. Certain sets of transfinite cardinality were referred to as "Cantor dust." Hermann Weyl (1885–1955), one of the greatest mathematicians of the twentieth century, used the expression "fog on fog" to derogatorily describe Cantor's ascending sequence of alephs [Stewart 96, p. 67]. Cantor himself was well aware of the brewing controversy and realized he was in opposition to views widely held, saying "I place myself in a certain opposition to widespread views on the mathematical infinite and to oft-defended opinions on the essence of number" [Kline 80, p. 200].

Chief among Cantor's critics was Leopold Kronecker, his former University of Berlin instructor. Kronecker's viewpoint was that of a finitist, and as such, he had no tolerance for mathematical objects which could not be constructed in a finite way. (He is well known for the statement, "God made the integers; all else is the work of man.") He carried these views to the extreme, going so far as to deny the existence of irrational numbers, thus reverting back to the Pythagorean philosophy of number. Kronecker strongly objected to Cantor's notions of infinity, referring to them as "mysticism." With both men being sensitive and temperamental, the attacks became personal. Kronecker referred to Cantor as a "charlatan" and "a corrupter of youth" [Dauben, p. 1] while Cantor blamed Kronecker directly for his failure in securing a professorship at the University of Berlin. An attempt at reconciliation was made by Cantor when he invited Kronecker to meet with him at a resort in the Harz Mountains. Kronecker accepted and the two met; yet, their differences would not be mended.

The mental pressures for Cantor mounted and his mental state began to collapse. Frustrated with his inability to solve the Continuum Hypothesis, depressed about his failure to achieve a position at the University of Berlin, and weary of his battles with Kronecker, he began to lose interest in formal mathematics, developing an irrational obsession with William Shakespeare and Francis Bacon. He promoted the unpopular hypothesis that Francis Bacon was the true author of Shakespearean plays and published, at his own expense, articles to this effect. It was as if Cantor's obsession with the Continuum Hypothesis had now been replaced with thoughts on the Bacon-Shakespeare connection. While delivering a lecture in Leipzig on the Bacon-Shakespeare hypothesis, he received word of the death of his youngest son, Rudolf. His mother died the same year.

In 1911, Cantor was invited to speak, as a Distinguished Foreign Scholar, at the University of St. Andrews, in Scotland. In that the offer had been extended by the university's department of mathematics, it was assumed that he would address topics in set theory and transfinite arithmetic. To the embarrassment of all, he spoke on the Bacon-Shakespeare connection.

During these troubled years Cantor would suffer three nervous breakdowns. His last years were spent in a mental institution in Halle. He died of a heart attack on January 6, 1918, having a funeral attended only by a few close family members.

The headstone reads

> **Dr. Georg Cantor**
> **Professor d. Mathematik**
> **3. 3. 1845–6. 1. 1918**

Despite the paradoxes and other difficulties associated with Cantor's work, it is recognized as monumental. Bertrand Russell referred to Cantor as one of the great intellects of the nineteenth century. David Hilbert described Cantor's transfinite arithmetic as "the most admirable flower of the mathematical intellect and one of the highest achievements of purely rational human activity" and strongly defended Cantor's infinity in saying, "No one shall drive us from the paradise Cantor has created for us" [Kline 80, p. 204].

In an attempt to salvage Cantor's work from its paradoxes and critics, set theory was axiomatized (formalized) in 1908 by the German mathematician Ernst Zermelo (1871–1953). He set up a system of eight axioms which defined the basic relations between sets and guaranteeing the existence of the empty set and infinite sets. The axiomatizing of set theory was analogous to Euclid axiomatizing geometry with his ten postulates. In 1922, the logician Abraham Fränkel (1891–1965) made additional contributions and the system (consisting of eight to ten axioms, depending on the formulation) is today known as Zermelo-Fränkel set theory.

1. The Axiom of Existence
2. The Axiom of Extension
3. The Axiom of Specification
4. The Axiom of Pairing

5. The Axiom of Unions
6. The Axiom of Powers
7. The Axiom of Infinity
8. The Axiom of Replacement
9. The Axiom of Choice

To the casual reader the axioms would seem reasonable enough, having been used implicitly by Cantor and others. (The first axiom guarantees the existence of at least one set. The second asserts that two sets are equal if and only if they have the same members. The other axioms appear equally plausible.) Zermelo merely formalized them. There is, however, a sleeper in the group. *Axiom 9—The Axiom of Choice* appears innocent. It states that if *S* is a collection of non-empty sets, then a new set of elements, called a *choice set*, can be formed by choosing one element from each set in the collection. If the collection of sets is finite, then the existence of a choice set is self evident, following in a finite number of steps from the other axioms. In such a case the Axiom of Choice, which bases the existence of a choice set on faith alone, is not required. Since the number of sets is finite (three in Figure 1.7), a choice set can actually be constructed.

Forming a choice set may be problematic if the collection of sets is infinite. Does such a choice set actually exist? In some cases it clearly does. If the collection were to consist of all closed intervals on the real number line of length one, then we could prescribe a method of choosing one number from each set, perhaps the midpoint, to form a choice set. That is, a choice set could consist of all midpoints of all closed intervals of length one. But what if the collection were the set of all subsets of real numbers? What method could one prescribe to choose from each set? There simply isn't one. All we can do is assume it can be done, and in making this assumption we invoke the Axiom of Choice.

Bertrand Russell clarified the meaning of this axiom with his well known *shoes and socks* example. To choose one shoe each from infinitely many pairs of shoes, a choice set could be formed by simply selecting the

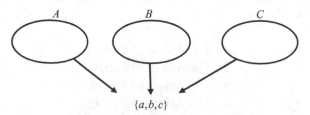

$\{a,b,c\}$

Figure 1.7. The Axiom of Choice.

right shoe from each pair. Each pair has a right shoe and thus a choice set is clearly defined. We do not merely *assume* we can do it. We've done it! On the other hand, to choose one sock each from infinitely many pairs of socks we must use the Axiom of Choice. There is no way to define, or exhibit a choice set. All we can do is assume its existence.

Russell noted [Vilenkin 68, p. 84], "At first it seems obvious, but the more you think about it, the stranger the deductions from this axiom seem to become; in the end you cease to understand what is meant by it."

Protests by such leading mathematicians as Emile Borel (1871–1956), Felix Bernstein (1878–1956), René Bair (1874–1932), Henri Lebesgue (1875–1941), and Jacques Hadamard (1865–1963), centered on the fact that a choice is being made, despite the fact no rule for the choice could be specified. How could such acts of faith be employed in mathematical proofs? Defenders of the axiom argue there would be no need to specify a rule to guarantee that a choice does exist. It is the *existence* that matters, not the rule itself.

As we see in the chapters to follow, assumption of the Axiom of Choice can lead to highly counterintuitive (and perhaps disturbing) results. The Banach-Tarski Theorem is just such an example. Some mathematicians prefer to avoid the axiom, since it is non-constructive and its use may have peculiar consequences, to say the least. Defenders argue that using the axiom yields mathematical results that could not be obtained otherwise. Furthermore, to deny the axiom would be more counterintuitive than the results obtained by using it. What now?

Gregory Moore notes [Moore 82, p. 2] the strange evolution of the Axiom of Choice as going from ". . . unconscious to conscious use and, for many mathematicians at the time, to conscious avoidance." Mathematicians avoiding the Axiom (and associated controversy) adopt the so-called ZF (axioms 1 through 8, no Axiom of Choice) axiomatic system. Mathematicians accepting the Axiom use the ZF + C = ZFC (axioms 1 through 9) axiomatic system.

But what are these highly counterintuitive and disturbing results that can follow from such an innocent appearing axiom? In *Introduction to the Foundations of Mathematics*, Raymond Wilder refers to the Axiom of Choice when he writes [Wilder 52, p. 74], "After all, who pays any regard to the blaze of a match until it starts a forest fire!" In 1924, Stefan Banach and Alfred Tarski lit the match and started a fire of controversy destined to burn until 1963 when Paul Cohen of Stanford University brought the issue to a bizarre conclusion.

Stefan Banach and Alfred Tarski

In 1924, the two Polish mathematicians Stefan Banach and Alfred Tarski published the thirty-four page paper entitled, "Sur la décomposition des ensembles de points en parties respectivement congruentes," translated as "On the decomposition of sets of points in respectively congruent parts."

Beyond their common nationality and 1924 collaboration, the two mathematicians had little in common, professionally or otherwise.

Stefan Banach (see Figure 1.8) was born March 30, 1892, in Cracow to Katarzyna Banach and Stefan Greczek. His parents were not married so it appears he was given his father's first name and mother's last name. Stefan never knew his mother as she gave him up after he was baptized; his father refused to reveal her identity. After being baptized he was sent to live with his paternal grandmother in Ostrowsko. When she took ill, he was sent by his father to live with Franciszka Płowa and her daughter, Maria, in Cracow. Maria's guardian, a French intellectual named Juliusz Mien, took a liking to Stefan, teaching him French and possibly igniting other academic interests, including mathematics.

Stefan's primary and secondary education was in Cracow where he became friends with Witold Wilkosz, a future professor of mathematics at Jagiellonian University, and Marian Albiński. Selections from Albiński's memoirs are given by Roman Kałuża in *Through a Reporter's Eyes—The Life of Stefan Banach* [Kałuża 96, p. 5]:

> Stefan Banach, as I remember him, was a good friend. Quiet, but not without a gentle sense of humor, he had a rather secretive nature. He always wore a clean and neat uniform, as we all did. His financial situation compelled him to tutor younger students as well as students *downtown*, but he never seemed needy. I should add that he tutored his own classmates without pay.

> From the early grades on Banach and Wilkosz were attached to each other through their love for mathematics. During school breaks I would often see them working on solutions to mathematical problems which, for me, a humanist, might as well have been Chinese.

Figure 1.8. Stefan Banach (1892–1945).
(Courtesy of the Mathematical Institute
of the Polish Academy of Sciences.)

Another classmate, Adolf Rożek recalled [Kałuża 96, p. 9]:

> Banach was slim and pale with blue eyes. He was pleasant in dealing
> with his colleagues, but outside of mathematics he was not interested
> in anything. If he spoke at all, he would speak very rapidly, as rapidly
> as he thought mathematically. He had such an incredible gift for fast
> thinking and computing that his interlocutors had the impression he
> was clairvoyant.

At the age of 18, Banach was off to Lvov Polytechnic to study
engineering. At that time Lvov was a beautiful multicultural city, a
center of Polish culture and science. It was there, in 1916, that he met
Hugo Steinhaus, a professor of mathematics at Lvov Polytechnic. The
chance meeting was significant both professionally and personally for
Banach as Steinhaus would be a strong influence for Banach in years
to come. Steinhaus had once claimed that Banach was the "greatest
discovery of his life" [Kałuża 96, p. 24]. It may be due to Steinhaus
that Banach chose to be a research mathematician; and, it was through
Steinhaus that Banach met and married Lucja Braus, who remained with
him for the rest of his life. Steinhaus and Banach started the journal
Studia Mathematica and the two mathematicians would jointly found
the Lvov School of Mathematics.

19

Prior to 1920, Banach was essentially self-taught, having no college (undergraduate) degree. For this and other reasons, he earned his doctorate in mathematics in a rather unconventional manner. Polish mathematician Otto Nikodym gives the following account [Kałuża 96, pp. 32–33] as to how Banach may have been tricked by colleagues into writing his thesis:

> . . . Lvov professors realized that the material for Banach's dissertation had been ready for some time, but that Banach had no intention of writing it down. Once Banach had proved his theorem, he was not very interested in turning it into a publishable paper. The process bored him. He was fascinated by mental speculations but abhorred the chores of putting them down neatly on paper. . . . Professor Ruziewicz instructed one of his assistants to accompany Banach on his frequent visits to the coffee houses, query him in a discrete fashion on his work, and afterwards write down Banach's theorems and proofs. When all of this information was typed out, the notes were presented to Banach, who then edited the text. This is how his PhD dissertation was finally completed.

Originally written in Polish, it was translated into French and published in *Fundamenta Mathematicae* as "Sur lés operations dans les ensembles abstraits et leur application aux équations integrals" ("On operations on abstract sets and their applications to integral equations"). The paper, regarded by many mathematicians as the birth of *functional analysis*, introduced *Banach spaces* and the *Banach fixed point theorem*, familiar concepts to many of today's research mathematicians.

Alfred Tarski (originally Alfred Teitelbaum) (see Figure 1.9), born January 14, 1902, in Warsaw, is regarded as one of the world's four greatest logicians, with Aristotle, Frege, and Gödel. Growing up in a financially secure family, he attended the Schola Mazowiecka, a school for intellectuals. Described by his teachers as having extraordinary ability (studying Russian, German, French, Greek, Latin, and Hebrew, besides the standard subjects), his plans were to study biology. After serving in the Polish army, Teitelbaum entered the University of Warsaw in 1918. Soon after taking a logic course, his logic professor, Stanislaw Lesniewski convinced him to switch his study to mathematics. He published his first paper, on set theory, when he was 19 years old and received his doctorate in 1924, being the youngest ever to be awarded this degree by the University of Warsaw. It was close to this time that

Figure 1.9. Alfred Tarski (1902–1983)—student photo at the University of Warsaw (circa 1918).

Alfred and his brother Wacław changed their religion from Jewish to Roman Catholic and changed their family name to Tarski. One speculates that anti-Semitism may have been a factor and these changes would make him appear more Polish, thus increasing his chances of a university appointment. It was also in 1924 that he and fellow Pole Stefan Banach published their famous paper.

With no doubt, it was Georg Cantor's groundbreaking achievements in set theory which paved the way for twentieth century mathematics. But it was a specific article, published by Felix Hausdorff in 1914, that actually triggered the collaboration by Banach and Tarski leading to the Banach-Tarski Theorem. Hausdorff's theorem, later known as the Hausdorff Paradox, asserts that a sphere (surface of a ball), minus a relatively small number of points, can be partitioned into three disjoint sets of points—A, B, and C—in such a manner that A, B, C, and the union of B and C are all congruent to each other. Details are given in Chapter 5 where we expose the paradoxes associated with this decomposition. In a sense, each of the three sets A, B, and C is both one-half and one-third of the sphere (minus the small number of points). It follows that the sphere, minus the small set of points, can then be decomposed into a finite number of sets and rearranged to form two copies of itself. Hausdorff's proof relies on the Axiom of Choice. (The proof of the Hausdorff Paradox is given in Chapter 5, as part of the proof of the Banach-Tarski Theorem.) Some mathematicians took

Hausdorff's result as evidence for rejection of the Axiom of Choice. Borel writes [Borel 14, pp. 255–256], "The contradiction has its origin in the application . . . of Zermelo's Axiom of Choice." Borel continues, "If one scorns precision and logic, one arrives at contradictions."

Hausdorff, however, gave a different interpretation, suggesting that some sets of points (in this case subsets of the sphere) had no definable measure of surface area, in which case there would be no paradox. Of course, the fact that some nonmeasurable sets could actually exist at all is paradoxical in itself.

This construction, by Hausdorff, is the heart of Banach and Tarski's paradox. Ten years after Hausdorff published his theorem, Banach and Tarski independently discovered a way to extend the paradox to the entire spherical surface, allowing for the duplication of the full sphere.

Sur la décomposition des ensembles de points en parties respectivement congruentes.

Par

St. Banach (Lwów) et A. Tarski (Varsovie).

Nous étudions dans cette Note les notions de *l'équivalence des ensembles de points par décomposition finie*, resp. *dénombrable*. Deux ensembles de points situés dans un espace métrique sont dits équivalents par décomposition finie (ou dénombrable), lorsqu'ils peuvent être décomposés en un nombre fini et égal (ou une infinité dénombrable) de parties disjointes respectivement congruentes.

Les principaux résultats contenus dans le présent article sont les suivants:

Dans un espace euclidien à n ⩾ 3 dimensions deux ensembles arbitraires, bornés et contenant des points intérieurs (p. ex. deux sphères à rayons différents), sont équivalents par décomposition finie.

Un théorème analogue subsiste pour les ensembles situés sur la surface d'une sphère; mais le théorème correspondant concernant l'espace euclidien à 1 ou 2 dimensions est faux.

D'autre part:

Dans un espace euclidien à n ⩾ 1 dimensions deux ensembles arbitraires (bornés ou non), contenant des points intérieures, sont équivalents par décomposition dénombrable.

La démonstration des théorèmes précédents s'appuie sur les résultats de MM. Hausdorff, Vitali et Banach[1]), qui concernent le problème général de mesure; elle fait donc usage de *l'axiome*

[1]) F. Hausdorff, *Grundzüge der Mengenlehre*, Leipzig 1914, p. 401 et 469.
G. Vitali, *Sul problema della misura dei gruppi di punti di una retta*, Bologna 1905.
St. Banach, *Sur le problème de mesure*, Fund. Math. IV, 1923, p. 30—31.

Figure 1.10. Banach and Tarski's paper [Banach and Tarski 24].
(Reprinted with the permission of *Fundamenta Mathematicae*.)

They then showed that the paradox could be further extended to the solid ball. Their co-authored paper (see Figure 1.10), written in French, appeared in *Fundamenta Mathematicae*, a respected Polish journal of mathematics. The authors chose French as it was more of an international language of scientific publication than Polish. The theorem now known as the Banach-Tarski Theorem or Banach-Tarski Paradox, states that a solid sphere (ball) can be decomposed into a finite number of pieces (as little as five) and reassembled in such a way as to form two solid spheres (balls) each identical in size to the original. This form of the theorem is referred to as the *duplication version*. An equivalent, and perhaps more striking, version asserts that a solid of any size and shape, say that of a small pea, can be decomposed into a finite number of pieces and reassembled to form a solid of any other size and shape, say that of the sun. This version, known as the *strong form* or *magnification version* of the theorem, is also known as the *pea and the sun* version, hence the title of this book. In Chapter 5 we first prove the duplication version, from which we easily establish the magnification version.

The fact that the theorem could be stated in simple, geometric terms, without mathematical jargon, allowed it to reach a large audience. It was controversial among mathematicians as its proof relied on the Axiom of Choice and its conclusion was indeed strange. But to the lay person, the conclusion itself attracted attention. Mathematicians can be tolerant, even appreciative, of counterintuitive results; but, the general public would be truly mystified by such a claim. J. W. Addison writes [Addison 83, p. 28] in an obituary for Alfred Tarski appearing in the *California Monthly*, "An irate citizen once demanded of the Illinois legislature that they outlaw the teaching of this result in Illinois schools!" (So it appears that Cantor wasn't the only mathematician attempting to corrupt young minds!)

So, the debates regarding set theory, the Axiom of Choice and now the bizarre theorem of Banach and Tarski intensified with no resolution to come for 40 years. It would take the work of Austrian mathematician Kurt Gödel and American mathematician Paul Cohen to ultimately bring down the curtain on this story.

Stefan Banach and Alfred Tarski would pursue their own interests after their singular collaboration. Banach continued writing papers with Steinhaus and, in 1939, Banach was elected President of the Polish Mathematical Society. Banach is credited with creating a *Polish style* of mathematics, where mathematicians would meet at coffee houses to socialize and work jointly to solve problems. He was a regular at Lvov's Szkocka Café (Scottish Cafe) where such mathematicians as Steinhaus, Ulam, Mazur, Kac, Schauder, and Kaczmarz would gather. At the time, Banach suggested that a record be kept of these problems, in the form of a notebook and held for safekeeping by the headwaiter of the cafe. An English version of the book entitled, *The Scottish Book: Mathematics from the Scottish Café*, edited by Daniel Mauldin with assistance from Jan Mycielski, has been published by Birkhäuser.

Banach was highly patriotic and remained in Poland, despite the wars and Poland's occupations by the Soviets and Germans. Kałuża gives the following account [Kałuża 96, p. 82] of Banach's attachment to Poland:

> . . . Von Neumann (1903–1957) an American mathematician of Hungarian descent, called by some "the Gauss of the twentieth century," visited Poland three times between the wars. Each time, on personal instructions from Norbert Wiener, father of cybernetics, he tried to talk Stefan Banach into emigrating to the United States; his last visit to Lvov took place in 1937. Responding to the latest job offer Banach asked:

> 'And how much is Professor Wiener willing to pay?'

> 'We anticipated this question', responded the confident American reaching into his pocket; 'here is a check signed by Professor Wiener on which entered only the numeral 1. Please add to it as many zeros as you deem fit!'

> Banach contemplated the offer for a moment and responded: 'This sum is too small to leave Poland' . . . .

In 1939, Soviet troops moved into Lvov. The Soviets, placing great value on Banach's contributions to mathematics, treated Banach well and appointed him to Dean of the Physical-Mathematical Faculty and Head of the Department of Mathematical Analysis at Jan Kazimierz University, renamed by the Soviets to Ivan Franko University.

In 1941 German troops entered Lvov and things took a turn for the worse for Banach and other Polish intellectuals. Intellectual life in Lvov was virtually destroyed with executions of Banach's friends and colleagues. Banach survived but conditions were grim. He was given the job as a feeder of lice in the Rudolf Weigl Bacteriological Institute which he kept for the remainder of the Nazi occupation of Lvov, until July, 1944.

Zdzisław Ruziewicz recalls [Kałuża 96, p. 89] the bizarre scene:

> ... The Institute employed everybody who had any direct or indirect contact with scholarly work, that is, a majority of the Lvov intelligentsia. I also worked at Weigl's Institute. We fed lice sitting at a long wooden table. I remember that almost instantly social groups began to form. There was a table where only humanities professors were feeders. Feeders who were natural scientists occupied another table. Banach and Knaster sat at the same table, and I had the impression that they were engrossed in mathematical conversations there.

Russian troops entered Lvov in 1944 and soon thereafter, under the Soviet/Polish government, Banach accepted a chair at the Jagiellonian University. He retained his position as President of the Polish Mathematical Society and was offered the position of Minister of Education. Unfortunately, his health deteriorated and he lost his battle with lung cancer in Lvov on August 31, 1945, at the age of 53.

Banach had earned respect as one of Poland's leading mathematicians and is considered by many a national hero (see Figure 1.11).

By contrast, Alfred Tarski's professional life was more international. He taught logic at the Polish Pedagogical Institute in Warsaw until 1925, and then taught mathematics at the University of Warsaw and Zeromski's Lycée in Warsaw until 1939. It was at the Lycée that he met and married Maria Witkowski, a teacher there.

Throughout the twenties and thirties Tarski did extensive work in mathematical logic, set theory, measure theory, and Boolean algebra. Tarski may be best known for the *Wahrheitsbegriff*, a work about the semantic conception of truth, which gives a mathematically rigorous definition of truth.

Tarski traveled extensively throughout Europe in the 1930s, visiting the University of Vienna and becoming part of the group of mathematicians, philosophers, and scientists of the Vienna Circle. Other members would include Kurt Gödel (whose work we will soon discuss) and Rudolf Carnap, who used Tarski's model of truth for his own work on semantics. In later years, ideas inspired by Tarski would be applied to semantics and computer programming.

Figure 1.11. Polish postal stamp.

By a twist of fate, Tarski came to the United States in 1939 to attend a meeting at Harvard University. Two weeks after his arrival, German armies invaded Poland. His Jewish ancestry would surely have been discovered had he remained in Germany (despite his name change and conversion to Catholicism), and his survival would have been unlikely. He petitioned to remain in the United States and, with the help of his colleagues, the petition was granted. His wife and two young children could not escape but they were able to survive and joined Tarski in the United States in 1946. Sadly, both his parents, his brother, and his sister-in-law were killed by the Nazis during the war.

After temporary appointments at Harvard, City College of New York, and Princeton, in 1942 he was appointed as a lecturer at the University of California, Berkeley and became a full professor there in 1946. He remained at Berkeley for the rest of his professional career, founding the Group in Logic and the Methodology of Science. Working in both the department of mathematics and philosophy, he encouraged recruitment of logicians, advocating that ten percent of the mathematics department be made up of logicians. His goal was achieved and he is directly responsible for turning the Berkeley campus into a world renowned center for mathematical logic.

Tarski journeyed from the Berkeley campus as a visiting professor to University College London (1950 and 1966), Henri Poincaré Institute in Paris (1955), the Miller Institute of Basic Research in Science (1958–1960), the University of California at Los Angeles (1967), and the Catholic University of Chile (1974–1975). His work is to be found in over 300 publications on topics including set theory, algebra, analysis, geometry mathematical logic, and semantics.

A. B. Feferman writes [Feferman 00]:

> A charismatic leader and teacher, known for his brilliantly precise yet
> suspenseful expository style, Tarski had intimidatingly high standards
> for students, but at the same time he could be very encouraging, and
> particularly so to women—in contrast to the general trend. Some
> students were frightened away, but a circle of disciples remained, many
> of whom became world-renowned leaders in the field.

His students included Andrzej Mostowski, Bjarni Jonsson, Julia
Robinson, Robert Vaught, C. C. Chang, Solomon Feferman, Richard
Montague, and Jerome Keisler, all of whom having achieved reputations
in mathematics and logic.

Tarski took teaching very seriously and he was highly respected by
his students. The following two anecdotes are provided by Steven R.
Givant [Givant 91, pp. 18–20] in his article "A Portrait of Alfred Tarski."
Givant was a student, research assistant, and close friend of Tarski at
Berkeley:

> During the 1969–70 academic year, the Berkeley campus was racked
> with student protests (mostly against the war in Viet Nam) that
> reached a climax when the national guard was called in and used tear
> gas to control the situation. Tarski's seminar that year was on the
> theory of relation algebras (a subject that had been revitalized through
> his own research and that was dear to him). There were around fifteen
> participants, of all political persuasions, and despite the general call
> to cancel classes, all wanted Tarski to continue teaching. Special
> arrangements were made for the class to meet off campus.

> The next year he lectured on the theory of general algebras. When
> it became apparent in early spring that he would not be able to cover
> his planned syllabus, the class asked him to talk an extra hour each
> meeting (for a total of 2 ½ hours). He was touched by their eagerness
> to learn and agreed to their requests, despite the extra work it meant
> for him. In later years he mentioned several times how much the
> request had moved him.

In 1981 he was awarded the Berkeley Citation, by the Regents of
the University of California. It is one of UC Berkeley's highest honors,
given to a selected few each year, chosen because of their achievement
in their fields and contributions to the university.

Steven Givant and Ralph McKenzie have produced and edited a four
volume collection of Tarski's papers entitled *Alfred Tarski—Collected*

Papers. In reviewing these volumes, John Corcoran writes [Corcoran 91, p. 4127]:

> The mathematical community owes a debt of gratitude to Givant and McKenzie for their efforts in producing this invaluable collection of Alfred Tarski's works. It is only when we see Tarski's papers collected in one place that we can begin to appreciate the scope and profundity of his influence on modern mathematical thought and, in particular, on modern mathematical logic. Mathematical logic as we know it today is almost inconceivable without Tarski's contributions.

Kurt Gödel – The Consistency of the Axiom of Choice

In 1931 a young Austrian mathematician named Kurt Gödel published a paper entitled, "Über formal unentscheidbare Sätze der Principia Mathematica und verwandter Systeme" ("On Formally Undecidable Propositions of Principia Mathematica and Related Systems"). The paper, appearing in *Monatshefte für Mathematik und Physik*, received little attention as only a small group of logicians and mathematicians were interested in the subject. It was a highly technical paper which was found unintelligible by many mathematicians. In 1932 he submitted the paper as his *Habilitationsschrift*, a qualifying paper required to enter the teaching profession. Within a few years, the mathematical world knew that "Gödel's Proof" had pulled the rug out from under twentieth-century mathematics. Gödel had shown that within an axiomatic mathematical system in which arithmetic can be developed there would be propositions that could neither be proved nor disproved. That is, we should expect some propositions to be *undecidable* as their verification is impossible within the given axiomatic system.

Ultimately, the work of Kurt Gödel and Paul Cohen would unravel the mysteries of Cantor's Continuum Hypothesis, the Axiom of Choice, and the Banach-Tarski Paradox itself!

Kurt Gödel (see Figure 1.12) was born April 28, 1906, in Brünn, Austria-Hungary (now Czechoslovakia) attending school there and completing high school in 1923. There were no scholars in Gödel's immediate family. His father is known to have had a trade school education. Much of what we know of the young Kurt Gödel is told by his older brother Rudolf [Casti and DePauli 00, p. 56]:

> Family life was harmonious. I got along very well with our brother, as did both of us with our parents. When he was around eight years old my brother had a bad case of rheumatism of the joints with a high

fever; since then he became a hypochondriac and imagined he had a heart defect, which was never medically established. In general, Kurt Gödel was a happy but shy child. He was very sensitive and was called "Herr Warum" (Mr. Why) because of his great curiosity.

He entered the University of Vienna in 1923 with intentions of studying physics, but soon switched to mathematics, completing his doctoral dissertation in mathematics in 1929.

Gödel was invited to join the Vienna Circle, bringing him in contact with Hans Hahn, Rudolf Carnap, Karl Menger, Bertrand Russell, and other significant philosophers, scientists, and mathematicians of the time. During the 1920s and 1930s, the group met weekly at the university to discuss the relationship between science and objective reality. The group's philosophy, *logical positivism*, was to become a leading philosophy of science through the 1950s.

Gödel's famous paper of 1931 contained two results which forced mathematics to undergo a sort of correction, ending the hope that all branches of mathematics could be described by a complete and consistent axiomatic system. His first theorem showed that any axiomatizable theory involving the natural number system is incomplete, in the sense that there would always be statements that could neither be proved nor disproved. Non-mathematical examples of such *undecidable statements* include self-referential statements such as, "This statement can not be proved." Gödel did not specify the exact axiomatic systems to which the theorem would apply, but it is clear that it applies to the Russell-Whitehead axioms for arithmetic and the Zermelo-Fränkel axioms (ZF and ZFC) for set theory.

Figure 1.12. Kurt Gödel (1906–1978).
(Photo courtesy of the Archives of the Institute for Advanced Study,
photographed by A. C. Wightman.)

His second theorem proves that for certain mathematical systems of axioms (involving the system of natural numbers), consistency is not provable in the system itself. In response, Hermann Weyl is known to have said that God exists because mathematics is undoubtedly consistent and the devil exists because we cannot prove the consistency. As Morris Kline writes [Kline 80, p. 263], "Gödel's result dealt a death blow to comprehensive axiomatization."

Though some would view his landmark result as an inadequacy or failure of mathematics, Gödel viewed it as a simple statement that mathematics and logic can not be completely mechanized and that human intuition and creativity would always be an essential component of mathematical progress.

How did Gödel obtain these landmark results? By assigning numbers to symbols, statements, and proofs, he showed that it was possible to arithmetize an axiomatic system. Each symbol, statement, and proof then had a *Gödel number* associated with it. Statements about the system (metamathematical statements) could also be assigned Gödel numbers. He then constructed an arithmetical assertion, he called it G, stating that the statement with Gödel number m is not provable. The statement is constructed so that its Gödel number is m. The self-referential nature of G makes it impossible to prove G is true. (Parallels to Gödel's theorems in physics, art, and music are given in this chapter's appendix.)

Rudy Rucker, in *Infinity and the Mind* [Rucker 82, p. 162], gives a non-mathematical version of Gödel's procedure in the form of a hypothetical interaction between Gödel and a machine, a Universal Truth Machine, capable of correctly answering any question submitted:

1. Someone introduces Gödel to *UTM*, a machine that is supposed to be a Universal Truth Machine, capable of correctly answering any question at all.

2. Gödel asks for the program and circuit diagrams of the *UTM*. The program may be complicated, but it can only be finitely long. Call the program *P(UTM)* for *Program of the Universal Truth Machine*.

3. Smiling a little, Gödel writes out the following sentence: "The machine constructed on the basis of the program *P(UTM)* will never say that this sentence is true." Call this sentence *G* for Gödel. *Note that G is equivalent to "UTM will never say G is true."*

4. Now Gödel laughs his high laugh and asks *UTM* whether *G* is true or not.

5. If *UTM* says *G* is true, then "*UTM* will never say *G* is true" is false. If "*UTM* will never say *G* is true" is false, then *G* is false (since *G* = "*UTM* will never say *G* is true.") So, if *UTM* says *G* is true, then *G* is in fact false, and *UTM* has made a false statement. So *UTM* will never say that *G* is true, since *UTM* makes only true statements.

6. We have established that *UTM* will never say *G* is true. So "*UTM* will never say *G* is true" is in fact a true sentence. So *G* is true (since *G* = "*UTM* will never say *G* is true.")

7. "I know a truth that *UTM* can never utter," Gödel says. "I know that *G* is true. *UTM* is not truly universal."

The reader may need to read this through again, and note that *G* is true in actuality, yet it can never be pronounced so by *UTM*. In more precise language, *G* is undecidable with respect to *UTM* and therefore *UTM* is incomplete.

In 1933 Gödel came to lecture at the Institute for Advanced Study at Princeton, New Jersey. He returned to Vienna at the end of the academic year with the intention of returning to Princeton; however, he suffered the first of a sequence of nervous breakdowns, delaying his trip back to Princeton until 1935. Soon after returning to Princeton he suffered a second nervous breakdown. He would return to Vienna and not lecture again until 1937.

Pictures of Gödel often show him as being thin and sickly in appearance. He was known to be a hypochondriac, with irrational fears of accidental (and in later years, intentional) poisoning. He was often malnourished and it is difficult to imagine how he could have gone through life caring for himself. He was most fortunate to have met and married Adele Porkert, a Viennese nightclub dancer, who would care for him until his death in 1978.

Adele was a Catholic divorcée, six years his senior, so they were indeed an odd couple. She was not approved of by Gödel's parents and Gödel's academic associates considered her to be no more than a prostitute. Gödel understood that a marriage would risk his career.

Nevertheless, the two were married in 1938 and the odd relationship of opposites would last, as they were highly devoted to each other. During his episodes of paranoia when he believed someone was trying to poison him, Adele would be his food taster.

Soon after their marriage, Gödel returned alone to the United States where he lectured at the Institute for Advanced Study and the University of Notre Dame. He returned to join his wife in Vienna in 1939 and was immediately called for military service by the Nazi armed forces. With the help of the Institute for Advanced Study, he was able to get exit visas for himself and his wife. The two arrived in Princeton in 1940 and Gödel would never leave the U.S. again.

In *The Consistency of the Axiom of Choice and of the Generalized Continuum Hypothesis with the Axioms of Set Theory* (1940), Gödel was able to show that the Axiom of Choice is consistent with the ZF system of axioms; that is, adding the Axiom to the ZF system (ZF + C = ZFC) would not lead to any contradictions. This is the same as saying that the Axiom of Choice can not be disproved. He did the same for the Continuum Hypothesis, thus leaving open the possibility of their independence from the ZF axiomatic system. To establish independence, it would be required to show that the Axiom of Choice and the Continuum Hypothesis could not be derived from the other ZF axioms. He tried but was unsuccessful in his attempts to do this. It would take the efforts of Paul Cohen of Stanford University to bring a sense of closure to these two open questions.

There are many stories about Gödel's idiosyncrasies. One involves Gödel's application for U.S. citizenship in 1948. In preparation for the citizenship exam, he studiously reviewed the U.S. Constitution. The day prior to the exam and interview, Gödel discovered what he perceived to be a flaw, or inconsistency in the Constitution which could theoretically allow the U.S. to be transformed into a dictatorship. Excited with his discovery, he contacted his friend, the noted economist Oskar Morganstern, telling him of his discovery. Morganstern and Albert Einstein were to serve as witnesses for Gödel at the scheduled interview and there was concern that Gödel could harm his own case by speaking of his discovery at the interview. He was cautioned to say nothing.

The next day, Einstein, Morgenstern, and Gödel drove to the federal courthouse in Trenton for the scheduled exam and interview. On the way to the courthouse, Einstein and Morgenstern did their best to keep Gödel focused on the exam and interview, not allowing his mind to wander. On his swearing in, he was asked by the judge of his opinion of the U.S. Constitution. Gödel could no longer contain himself, launching into a discourse on the Constitution's inconsistencies. Reports have it that it took the efforts of the three— Morganstern, Einstein, and the judge—to calm Gödel down and get on with the rest of the interview. Gödel passed the exam and was granted citizenship.

Gödel and Einstein were close friends at the IAS. There is the story of Gödel seeing a picture of General MacArthur on the front page of the New York Times. He told Einstein that the picture in the paper was that of an imposter. When asked to explain he told Einstein that he had an older picture of MacArthur and the ratio of the nose length to the distance from the tip of the nose to the point of the chin differed on the two pictures!

Gödel was highly reclusive and notorious for not keeping scheduled appointments. When asked why he bothered to make appointments if he had no intention of keeping them, he replied that it was his way of guaranteeing that he would not have to meet his visitor.

Gödel's last published paper was in 1958, after which he became increasingly ill, both mentally and physically. His wife Adele suffered a debilitating stroke and his good friend Oskar Morganstern died of cancer. On January 14, 1978, Gödel died in the Princeton Hospital of "malnutrition and inanition (exhaustion and a general lack of strength) caused by personality disturbance." His physician, Dr. Ramona, stated [Caste and DePauli 00, p. 92], "He had refused all food. He had never eaten very much, but his final weight was only around sixty pounds. He died in the fetal position." Effectively, he starved himself to death.

Though he published fewer papers than any other of the world's great mathematicians, except Bernhard Riemann, Gödel's work has been considered revolutionary in logic and the foundations of mathematics. He is described as one of mathematics' greatest Platonists, as shown by comments about mathematical sets [Barrow 92, p. 261]:

> . . . despite their remoteness from sense experience, we do have something like a perception also of the objects of set theory, as is seen from the fact that the axioms force themselves upon us as being true. I don't see any reason why we should have less confidence in this kind of perception, i.e., in mathematical intuition, than in

sense perception, which induces us to build up physical theories and to expect that future sense perceptions will agree with them and, moreover, to believe that a question not decidable now has meaning and may be decided in the future. The set-theoretical paradoxes are hardly any more troublesome for mathematics than deceptions of the senses are for physics.

In 1965, Oskar Morganstern wrote to Austria's Foreign Minister Bruno Kreisky [Casti and DePauli 00, p. 3]:

> There is absolutely no doubt that Gödel is the world's greatest living logician . . . Einstein once told me that his own work no longer meant much to him, and that he simply came to the Institute to have the privilege of walking home with Gödel.

Paul Cohen – The Independence of the Axiom of Choice

American mathematician Paul Cohen (see Figure 1.13) distinguished himself by solving problem number one of Hilbert's famous list of twenty-three unsolved problems. In doing so, he settled the questions associated with Cantor's Continuum Hypothesis and Zermelo's Axiom of Choice. In that the proof of the Banach-Tarski Theorem hinges on the Axiom of Choice, Paul Cohen will be the final act in our history of the Banach-Tarski Theorem. He has been honored with two of the most prestigious awards in mathematics—the Fields Medal and the Bôcher Prize.

Born April 2, 1934, in Long Branch, New Jersey, Cohen grew up in Brooklyn as the youngest of four children born to Jewish immigrant parents. He became interested in mathematics at nine years of age when his older sister Sylvia would show him problems from her algebra book. He would go to the Brooklyn Public Library, reading all he could of the mathematics books available. At ten years of age he was able to solve quadratic and cubic equations and had begun reading books on the theory of equations.

Being so mathematically talented at that age made it difficult for him to get help with mathematical questions. The library did not allow children to browse the adult books, so he had to sneak into the mathematics section. And, he had no one to turn to with questions. As Cohen describes it [Albers, Alexanderson, and Reid 90, pp. 45–46]:

Figure 1.13. Paul Cohen (1934–).
(Courtesy of Paul J. Cohen.)

I couldn't turn to anyone in my family for that kind of help. They didn't discourage me, but they didn't know what to make of me. There were a few people in the neighborhood studying engineering or science who gave me some books after they heard I was reading math on my own. And Sylvia continued to get me books from the college library . . .

To summarize, by the time I was in sixth grade I understood algebra and geometry fairly well. I knew the rudiments of calculus and a smattering of number theory, which I liked very much. I felt rather isolated. A lot of teachers are very threatened when they find a child is studying advanced things. And I was reluctant at that time to talk to other children because I felt they found my interest in math somewhat strange.

At Stuyvesant High School in lower Manhattan, Cohen was able to meet other students with interests similar to his own. Stuyvesant was one of several schools in the area specializing in science and mathematics; so, the previous isolation he felt began to reside.

It was there he decided to become a mathematician. With his family's financial situation as a consideration, Cohen attended college locally at Brooklyn College from 1950 to 1953. Some of his high school friends had gone to the University of Chicago and encouraged him to apply there; he did and was accepted. Cohen recalls [Albers, Alexanderson, and Reid 90, pp. 45–46]:

> I studied seriously there (Brooklyn College) but I felt bored. . . . Also, I didn't enjoy the routine of all the required courses. I really wanted to do math. . . . It (the University of Chicago) was a very exciting and innovative place at that time and would take people for graduate study who had just two years of college.
>
> . . . one of my first memories of Chicago was walking down the corridor of Eckhart Hall and hearing him (Irving Kaplansky) say that he had just proved that every B* algebra is a C* algebra—he was handing out mimeographed sheets in the hall. I thought 'Boy, you know this is where mathematics is being made—a B* algebra is a C* algebra!' Then I thought, 'What is a B* algebra?' I knew then that I was at a topflight school!

At Chicago he earned his MS degree in 1954 and PhD degree in 1958 with a doctoral thesis entitled *Topics in the Theory of Uniqueness of Trigonometric Series.* Prior to earning his PhD, he held a one year instructor appointment at the University of Rochester. It was during these years at Chicago while working on a problem in number theory that Cohen first considered Gödel's Incompleteness Theorem. From childhood Cohen had been interested in number theory and during the years at Chicago he was attempting to solve a specific number theory problem, with no success. A graduate student suggested to Cohen, "You certainly can't get a decision procedure for even such a limited class of problems, because that would contradict Gödel's theorem." Cohen read the theorem, at first finding it more philosophical than mathematical. He soon realized that Gödel's theorem did apply to the problem he had been considering. Though convinced of the theorem's applicability, he had no interest in it.

With regard to Hilbert's first problem of Cantor's Continuum Hypothesis and the Axiom of Choice, Cohen recalls [Albers, Alexanderson, and Reid 90, p. 51]:

> I have this memory of myself standing in the library at Chicago and looking at a list of famous problems—perhaps it was even Hilbert's list. It included the problem of the Continuum Hypothesis, and I said to myself, 'That's one problem that I really don't understand.' I wondered what would it actually mean to solve it. I had a

feeling that it really wasn't a problem in the same sense as other mathematical problems, but I definitely thought it was something that I wasn't interested in.

After receiving his PhD, Cohen spend one year at the Massachusetts Institute of Technology, followed by two years as a fellow at the Institute for Advanced Study at Princeton. Curiously, there is no evidence that he and Gödel met during their coincident employment at the IAS. In 1961 he joined the faculty at Stanford University where he remains today, with interests in harmonic analysis, differential equations, and set theory.

While at Stanford, Cohen continued the work begun by Gödel, by showing that the Continuum Hypothesis and the Axiom of Choice are independent of the ZF axioms of set theory. Gödel had accomplished part of the job by showing that if the ZF system is consistent without the Continuum Hypothesis, then the system would remain consistent by adjoining the Continuum Hypothesis. He did the same with the Axiom of Choice. So, Gödel had shown that the Continuum Hypothesis and the Axiom of Choice could not be disproved. Gödel tried without success to show that the Continuum Hypothesis and the Axiom of Choice could not be derived from the ZF system. If someone could do this, then both open questions would be declared undecidable.

It was Solomon Feferman, Professor of Mathematics and Philosophy at Stanford University, that suggested Cohen have a look at these problems. Initially Cohen felt that the problem was not well stated and that one would have had to be "slightly crazy even to think about the problem" [Albers, Alexanderson, and Reid 90, p. 52].

One may speculate if being *slightly crazy* is a prerequisite or consequence of working on this problem. Georg Cantor spent the last years of his life in a mental institution in Halle, suffering depression brought about by the death of his youngest son, frustrations with his inability to solve the Continuum Hypothesis and his ongoing battles with Leopold Kronecker. Ernst Zermelo, known for formalizing set theory and the Axiom of Choice, had at least one mental breakdown. And Kurt Gödel's battles with depression and paranoia sent him into depths of illness similar to that of Georg Cantor. Are there links between the nature of their work and their mental illness?

Cohen was more interested in solving the Axiom of Choice than the Continuum Hypothesis simply because he believed it to be more significant. He felt the Continuum Hypothesis to be of more philosophical interest, whereas the Axiom of Choice was used more in proofs. He worked on both, making some progress, but became frustrated and in 1962 stopped work for several months. While on a road trip through the Southwest with his future wife he began thinking about the problem again, becoming convinced that it could be done.

In April, 1963, Cohen was convinced he had a solution. He had invented a technique called *forcing*, where certain set theoretic models can be extended in such a way so that a given statement (or negation of a statement) becomes true (is forced to be true) in the extended model. Using this technique, Cohen showed that the ZF system would remain consistent if the Axiom of Choice and the Continuum Hypothesis did not hold. That is, he showed that the Axiom of Choice and the Continuum Hypothesis could not be derived from the ZF axioms. This, combined with Gödel's results, established the independence (undecidability) of these issues within this axiomatic system. The technique was both new and powerful and it seemed as if many of the open questions of set theory could be solved by this method. With extreme and understandable excitement he hand carried a draft of his work to Gödel at Princeton.

In June of that year Cohen received Gödel's response. Gödel wrote [Yandel 02, p. 78]:

> I hope you are not under some nervous strain which hampers you in your work. You have just achieved the most important progress in set theory since its axiomitization. So you have every reason to be in high spirits.

The Stanford campus is only fifty miles from the Berkeley campus of the University of California. Alfred Tarski and the other logicians at Berkeley were well aware of what Cohen had accomplished, even before it was published. Tarski remarked [Yandel 02, p. 81], "They have a new method and they'll get everything."

Cohen describes his own proof [Albers, Alexanderson, and Reid 90, p. 58]:

> One might say in a humorous way that the attitude toward my proof was as follows. When it was first presented, some people thought it was wrong. Then it was thought to be extremely complicated. Then it was thought to be easy. But of course it is easy in the sense that there is a clear philosophical idea. There were technical points, you know,

which bothered me, but basically it was not really an enormously involved combinatorial problem; it was a philosophical idea.

Gödel wrote of the undecidability of Cantor's Continuum Hypothesis, clearly portraying his Platonistic interpretation [Barrow 92, p. 264]:

> Only someone who denies that the concepts and axioms of classical set theory have any meaning . . . could be satisfied with such a solution, not someone who believes them to describe some well-determined reality. For in reality Cantor's conjecture must be either true or false, and its undecidability from the axioms as known today can only mean that these axioms do not contain a complete description of reality.

Cohen suggests otherwise [Albers, Alexanderson, and Reid 90, p. 54]:

> Well, philosophically, if you really believe sets exist—I mean, if you adopt the extreme Platonic position—you can ask, what is the answer? Certainly Gödel himself had the Platonic view that the question demanded an absolute answer and that, therefore, neither his proof of the consistency of the Continuum Hypothesis with the axioms of set theory nor mine of its independence from them was a final answer. My personal view is that I regard the present solution of the problem as very satisfactory. I think that it is the only possible solution. It gives one a feeling for what's impossible, and in that sense I feel that one should be very satisfied. There are further problems, but they are fairly technical ones. If I were a betting man, I'd bet no one is going to come up with any other kind of solution. There will be philosophical papers, but I don't think any mathematical paper will say that there is any answer other than the answer that it's undecidable.

Cohen's method has since been modified and has been applied to problems outside of set theory, including infinite graphs, measure theory, and infinite games. He was awarded the Fields Medal at the International Congress of Mathematicians in Moscow in 1966 for specific work in the foundations of set theory. The award, named after the Canadian mathematician John Charles Fields, is often referred to as the *Nobel Prize of Mathematics*. In that the award's intent is to give recognition and support to mathematicians showing promise of future achievement, it may only be awarded to mathematicians under the age of forty. In Cohen's case, it was the first and only time the award was given for work in logic and foundations.

Proof of the Banach-Tarski Theorem, within the ZFC system, relies on the Axiom of Choice. So is the Banach-Tarski Theorem true? It is, if we assume the Axiom of Choice is true. Is the Axiom of Choice true? It may be, but we must accept the fact that we can not decide on the basis of the ZFC system. It is in every sense a *matter of choice*!

The work of Gödel and Cohen in showing the limitations inherent to such axiomatic systems may appear to hinder mathematical growth, exposing pre-existing roadblocks on the road to discovery. In actuality, it shed an unexpected light on set theory and opened new paths to explore. To best understand why, we can look to Euclidean geometry and the question of Euclid's *fifth postulate*. Sometime around 300 BC Euclid axiomatized geometry by establishing a set of postulates (axioms) from which all other properties could be proven as theorems. This may have been an attempt to resolve the paradoxes of Zeno, much like Zermelo's axiomitization was an attempt to resolve the paradoxes of set theory. The so called *fifth postulate* or *parallel axiom* states that through a given point there can be only one line parallel to a given line. Seems reasonable! The problem with this axiom is that there is no way to verify it; it is not self-evident. Parallel lines, extending infinitely far, are not directly observable. So, one must take the axiom on faith, much like one takes the Axiom of Choice on faith for set theory.

Nineteenth-century mathematicians Carl Friedrich Gauss, Johann Bolyai, and Nicolai Lobachevsky independently worked on the possibility of *non-Euclidean geometries* for which the parallel postulate did not apply. In 1840 Lobachevsky was the first to publish a systematic non-Euclidean geometry—*Lobachevskian geometry*. Bernhard Riemann (1826–1866) proposed a large variety of non-Euclidean geometries, known as *Riemannian geometries*. As non-Euclidean geometries were proposed, two questions arose. First, could there be a consistent non-Euclidean geometry? That is, would a system of axioms without the parallel postulate as stated by Euclid be consistent and not lead to contradictions? If so, would there be any applicability associated with a non-Euclidean geometry? The motivation for the second question should be clear to the reader unfamiliar with non-Euclidean geometries. Euclidean geometry as learned in high school may appear as the only geometry of physical reality. Could some other geometry accurately describe our universe? Why not? The Euclidean geometry of the plane may not be the best geometry for describing curved surfaces, like that of the earth. If one stops to think about it, is a curved surface, like that of the earth, any less natural than a flat surface as used by Euclid?

The first proof of the consistency of a non-Euclidean geometry was by Eugenio Beltrami (1835–1900), when in 1869 he showed that the non-Euclidean geometry of Lobachevsky and Bolyai was consistent. Once the parallel postulate was shown to be independent of the other postulates of Euclidean geometry, it encouraged the creation of other geometries—non-Euclidean geometries, which, by definition, do not assume the parallel postulate. One then has a choice of geometries with which to solve problems, mathematical and physical. In 1915 Einstein based his theory of gravitation on the assumption of a non-Euclidean geometrical description of the universe.

Today there are many classes of non-Euclidean geometries, each consistent and applicable to a specific type of problem. *Spherical (Riemannian) geometry* is a non-Euclidean geometry defined on the surface of a sphere where straight lines are great circles and the distance between two points on the sphere is measured along the arc of the great circle joining the two points. The parallel postulate does not apply because there are no parallel lines. *Taxi-cab geometry*, where distances are measured as sums of horizontal and vertical distances along a grid imposed on the space, is both recreational and useful. This chapter's appendix contains a *hyperbolic geometry* (non-Euclidean) model of the Banach-Tarski Theorem, in which the Axiom of Choice is not required to produce the paradox.

Similarly, independence of the Axiom of Choice allows a choice of axiomatic systems for set theory. Axiomatic systems, known as non-standard or non-Cantorian systems, are being developed without the Axiom of Choice. If Hilbert *directed* with his famous list of twenty-three unsolved problems, and Gödel *corrected* by having mathematics adjust itself self-referentially, then Cohen *re-directed* by giving twentieth century mathematics a powerful new tool in that of forcing. With independence should come a sense of freedom and optimism for twenty-first century mathematics. In the words of Cantor, "The essence of mathematics lies in its freedom."

Appendix

Examples of Self-Reference in Physics, Art, and Music

An interesting parallel to Gödel's theorem in the physical world was given by the Nobel Prize winning German physicist Werner Heisenberg (1901–1976). In 1927, as part of the theory to quantum mechanics, he formulated the *Uncertainty Principle*, stating that it is impossible to simultaneously specify the position and momentum of a physical object, regardless of the accuracy of the devices used to make these measurements.

To see the analogy with Gödel's theorem, we must back up to the beginning of the nineteenth century when the French scientist Pierre-Simon Laplace (1749–1827) proposed the doctrine of *scientific determinism*. The idea was that the future of the universe would be, in theory, completely predictable if we knew the complete and exact state of the universe at a given time. For example, if we knew the exact position and momentum of the sun and planets of our solar system at some point in time, then the mathematical calculations of Newtonian mechanics could be used to predict, with perfect accuracy, the exact nature of the solar system at any specified time in the future. It was pure mechanics, much like predicting the future position of billiard balls on a table, given their present positions and velocities. Laplace went so far as to suggest similar laws would govern human behavior, thus arguing against free will. There were objections to the deterministic doctrine by those believing in God's ability to intervene and an individual's free will.

Heisenberg noted that to predict the future position of say, an atomic particle, its present position and velocity would have to be measured with perfect accuracy. But to do so, would involve disturbing the particle itself, creating a self-referential difficulty of *measuring the effects of the measurement*. For example, to accurately measure the position, a short wavelength (high energy) of light would be required, disturbing the velocity by a significant amount. It is a bit like *robbing Peter to pay Paul*. Similarly, to measure velocity accurately, one must sacrifice the accuracy of position. The same kinds of issues arise in conducting public opinion surveys. Is it possible to accurately solicit a person's opinion without affecting, in some way, his response?

When first introduced by Heisenberg, the principle was met with some objection as it seemed to contradict the fundamental laws of cause and

Heisenberg may have slept here!

Figure 1.14. Roadside attraction.

effect. Heisenberg's Uncertainty Principle along with quantum mechanics was the death blow to scientific determinism in that there could be no precise prediction of a future state of a system if there was no hope in making an accurate measurement of the present state of the system.

Gödel's theorem did for mathematics what Heisenberg's Uncertainty Principle did for physics. Prior to Gödel's 1931 theorem, the formalism movement of mathematics, lead by David Hilbert, proposed that all of mathematics could be established algorithmically in a consistent and complete way, thus reducing mathematics to a mechanical and deterministic system.

In his lecture delivered to the International Congress of Mathematics in 1900, Hilbert declared [Yandell 02, p. 395]:

> However unapproachable these problems may seem to us and however helpless we stand before them, we have, nevertheless, the firm conviction that their solution must follow by a finite number of purely logical processes. . . . This conviction of solvability of every mathematical problem is a powerful incentive to the worker. We hear within us the perpetual call: There is the problem. Seek its solution. You can find it by pure reason, for in mathematics there is no *ignorabimus*.

Gödel's theorem shattered Hilbert's vision.

Self-referential paradoxes appear in art as well. The Dutch artist Maurits C. Escher (1898–1972) captures the concept in several popular works. The lithograph *Three Spheres II* (1946) depicts three spheres, the middle of which mirrors the outside two, as well as the artist sketching

Figure 1.15. M. C. Escher's Print Gallery.

all three. The lithograph *Drawing Hands* (1948) depicts a pair of hands, the right drawing the left, which draws the right drawing the left, . . . *Print Gallery* (1956) is a detailed lithograph depicting a young man looking at prints on a wall, one of which shows a woman looking out her window down to a sloping roof under which the gallery itself is housed. So the young man views a picture of which he is a subject. Reality and image are one and the same (see Figure 1.15).

Johann Sebastian Bach shows self-referential themes in his canons and fugues. In the last line of *Art of Fugue*, he encodes his own surname *B-A-C-H* (See Figure 1.16). (In German, B natural is designated as H.)

Sadly, this was the last piece of music that he wrote before his death.

B   A   C   H

Figure 1.16. Last line of Bach's *Art of Fugue.*

Douglas R. Hofstadter, in his 1980 Pulitzer Prize winning book, *Gödel, Escher, Bach: An Eternal Golden Braid*, refers to such self-referential phenomena as *strange loops.* According to Hofstadter [Hofstadter 79, pp. P-2–P-3]:

> . . . the strange loop notion holds the key to unraveling the mystery that we conscious beings call 'being' or 'consciousness'. . . . The Gödelian strange loop that arises in formal systems in mathematics (i.e., collections of rules for churning out an endless series of mathematical truths solely by mechanical symbol-shunting without any regard to meanings or ideas hidden in the shapes being manipulated) is a loop that allows such a system to *perceive itself,* to talk about itself, to become *self-aware,* and in a sense it would not be going too far to say that by virtue of having such a loop, a formal system *acquires a self.*

A Non-Euclidean Model of the Banach-Tarski Paradox

The Banach-Tarski Paradox does not hold in the plane; a space of three or more dimensions is required. However, Stan Wagon and Jan Mycielski give us a geometrical interpretation of the Banach-Tarski Paradox (or more accurately the Hausdorff Paradox) in hyperbolic (non-Euclidean) space [Wagon 93a]. The sets are constructible and can be visualized in the hyperbolic plane. Even more surprising, the Axiom of Choice is not required!

Wagon and Mycielski use the *Poincaré disc* model of the hyperbolic plane. Points of the plane are represented by the interior points of a disc. The circular boundary of the disc represents *points at infinity.* Straight lines are represented either as diameters drawn through the disc's center or as circular arcs intersecting the circular boundary at right angles. We view the hyperbolic plane from two different points of view, or *portholes* [Dewdney 89, p. 117]. Figure 1.17 shows the hyperbolic plane partitioned into *triangular* regions, which can be grouped to form three subsets of the hyperbolic plane—*A, B,* and *C* (see Figure 1.17).

Chapter One

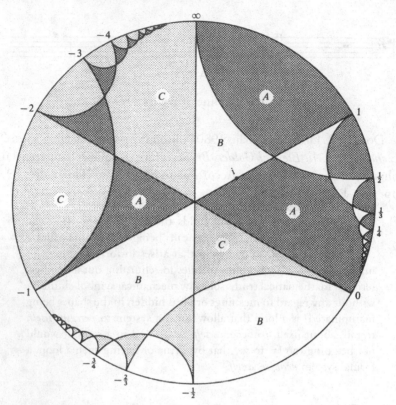

Figure 1.17. The Poincaré disc model of the hyperbolic plane. Region A (dark grey) is ⅓ of the entire space [Wagon 85, Figure 5.2(a)].
(Reprinted with the permission of Cambridge University Press.)

Consider first region A of the disc. If the disc is rotated 120° clockwise, then A will coincide exactly with B. If the disc is rotated an additional 120° clockwise, then A will coincide with C. So, as viewed through this porthole, all three pieces are congruent, each making up one-third of the hyperbolic space.

But now look at Figure 1.18 which views the same hyperbolic space through a different porthole, centered at point *i*. The same region A (dark grey and appearing differently as viewed differently) is congruent to the two other areas combined. A 180° rotation of the disc below should make this clear. So from this viewpoint, A corresponds to one-half of the space.

Summarizing, the region A is congruent to one-third of the hyperbolic space and one-half of the hyperbolic space, indicating a two dimensional non-Euclidean version of the Hausdorff Paradox. And, it was done without the Axiom of Choice!

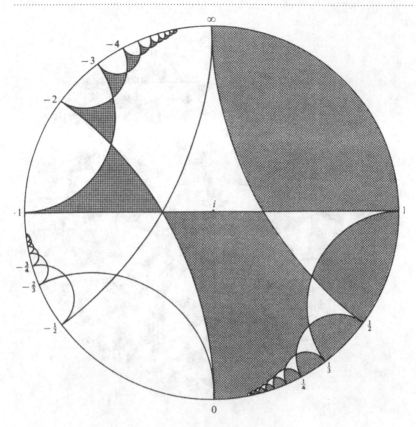

Figure 1.18. Center of model shifted to point i. Region A (dark grey) is ½ of the entire space [Wagon 85, Figure 5.2(b)].
(Reprinted with the permission of Cambridge University Press.)

In addition to his artistic impressions of self-reference, Escher uses the Poincaré disc in four of his works—*Circle Limit I*, *Circle Limit II*, *Circle Limit III*, and *Circle Limit IV*. In the woodcut *Circle Limit IV* (1960) (see Figure 1.19), interlocking angels and devils tessellate the disc becoming infinitesimally small near the disc's circumference at infinity. It represents the balance of good and evil (heaven and hell) in the world.

In 1993 Stan Wagon suggested the possibility of the Circle Limit IV woodcut being used to model the Banach-Tarski Theorem in a manner similar to Wagon's Poincaré disc model. Curtis Bennett, currently associate professor of mathematics at Loyola Marymount University, provides such a model in his article, "A Paradoxical View of Escher's

47

Figure 1.19. M. C. Escher's Circle Limit IV.

Angels and Devils" [Bennett 00]. The eight page article, appearing in *The Mathematical Intelligencer*, artistically illustrates the paradox by using two unique coloring schemes for the Escher figure.

2 Jigsaw Fallacies and Other Curiosities

*I ordered the "two piece dinner" at El Pollo Loco but returned
to the counter with my order when I noticed one piece of chicken
on my plate instead of two. I tactfully pointed out the error,
showing my receipt. The cashier acknowledged the error,
giving my plate back to the cook at the grill. He shrugged, and
with one drop of the cleaver, split the single piece of chicken in
two. They returned the plate to me without a second thought.
I guess it all depends how you look at it!*

The Banach-Tarski Theorem can be called a three dimensional jigsaw
paradox in that a partitioned solid sphere can be reassembled to form
two copies of itself. Or, the pieces can be reassembled in such a fashion
as to form a single sphere many times the size of the original. The idea
of getting something for nothing must intrigue us all. Reversing the
process is equally magical as it represents a disappearing act. When this
happens before our eyes we suspect some sort of deception, whether it
be done in grand scale as stage magic, close up at the personal level, or
in the sometimes physical, sometimes theoretical world of mathematics.
It is natural for us to ask, "How is this done?"

Paradoxes and Antinomies

The word *paradox*, coming from *para* and *doxos*, translates literally as
beyond belief.

Nicholas Falletta, in his book *The Paradoxicon*, refers to a paradox
[Falletta 83, p. xviii] as "truth standing on its head to attract attention."
Thus, paradox often refers to appearance, requiring an explanation.

Things appear paradoxical, perhaps because we don't understand them, perhaps for other reasons.

Painting with a broad stroke, paradoxical statements, or arguments, can be one of three types.

Type 1.   A statement which appears contradictory, even absurd, but may be true in fact. (The Banach-Tarski Theorem is of this category, as the conclusion of the theorem appears to contradict common sense; yet, the conclusion is true. As a prelude to the Banach-Tarski Theorem, the "Baby BTs," given in Chapter 4, are also Type 1 paradoxes.)

Type 2.   A statement which appears true, may be self-contradictory in fact, and hence false. These are fallacies, following from a fallacious argument.

Type 3.   A statement may lead to contradictory conclusions. This is known as an *antinomy*.

Whichever the category, there is an element of surprise associated with the paradox and the reader may feel compelled to resolve matters. In this chapter, examples of all three types are given, though none is as stunning as that of Banach and Tarski, which is explained in Chapter 5.

The Type 1 paradox, true in actual fact, is often resolved with additional information and reconsideration. For example, data will show that U.S. cities having the greatest number of churches also have the highest homicide rates (measured in homicides per year). Though paradoxical, there is a simple explanation. The number of churches and the number of homicides are each roughly proportional to the city's population. Large cities have large numbers of each; the opposite would be true for small towns.

A fascinating Type 1 physical paradox has been invented by Joel Cohen of Rockefeller University. He uses the spring and string contraption, shown in Figure 2.1, to model *Braess's Paradox*, a phenomenon relating to the study of crowded networks such as traffic congestion. It is presented here as a physical paradox (Type 1) without reference to networks. A thorough discussion of Braess's Paradox is given in the article "Road to Ruin" [Bass and Martin 92], by T. Bass and A. Martin, appearing in the May 1992 issue of *Discovery Magazine*.

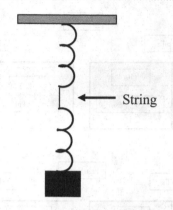

Figure 2.1. Hanging weight.

A weight is hung from a spring. The top of the spring is then attached to the bottom of an identical spring by means of a small piece of string. The top of the second spring is then attached to the ceiling.

Clearly, if the small linking piece of string is cut, the weight will fall straight down.

To prevent the weight from falling all the way to the floor, imagine two additional lengths of string are attached, as shown in Figure 2.2. These pieces are of equal length, longer than the short piece of linking string. These longer pieces are hanging loose, not taut as is the small linking piece. (See example on the left in Figure 2.2.)

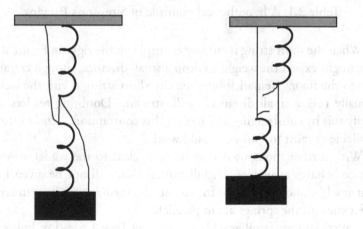

Figure 2.2. Hanging weight with additional strings—before and after cut.

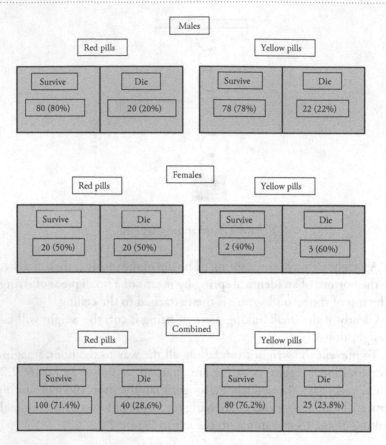

Table 2.1. A hypothetical example of Simpson's Paradox.

When the short string is cut (see example on the right in Figure 2.2), one might expect the weight to drop a small distance, though certainly not to the floor. Remarkably, when the short string is cut, the weight actually rises a small distance, as illustrated. Doubting readers may verify this by constructing and testing this contraption. *Slinky* springs, available at most toy stores, should work.

With further thought, it may become clear to the reader why the device behaves as it does. A full explanation will not be given here. But as a hint, note that prior to the cut, the springs are acting in series. After the cut, the springs act in parallel.

Statisticians are familiar with a surprising Type 1 paradox, known as *Simpson's Paradox*, which can occur when comparing success rates (or incident rates) for two populations.

A hypothetical example will be presented with real world examples given in this chapter's appendix.

A researcher wishes to compare the effectiveness of two medications (red pills vs. yellow pills) on a rare malady which, if not treated, will be fatal. Because males and females may respond differently to these treatments, data is recorded for each sex. In total, 245 patients (200 males and 45 females) are treated with each patient being given one of the two medications. A summary of the findings is given in Table 2.1. Assume the data is typical of all patients to be treated with these medications. There are no aberrations.

Note that for males, the red pill appears to have a better success rate than the yellow pill (80% > 78%). For females, the red pill performed better than the yellow pill (50% > 40%). So, whether male or female, it would appear that the red pill would be the treatment of choice.

Our researcher passes the data to a colleague who pools the data, disregarding sex. As the combined data shows, patients given the yellow pills had a significantly higher survival rate (76.2%) than those given the red pills (71.4%). So overall, one could argue that yellow pills should be the treatment of choice. This represents a complete reversal of the findings for each sex where the red pills were found preferential.

To further illuminate the paradox, the reader should try to answer the following questions:

1. If you were the patient, which treatment (red or yellow) would you prefer?
2. Assume you are a physician at the clinic and have been notified of an incoming patient requiring treatment. You have no idea if the patient is male or female. If these medications require preparation, which color pill would you prepare?

Once the reader thinks this through, the paradox should disappear and the answers to the above questions should become clear. An explanation is given in this chapter's appendix.

A Type 2 paradox differs from a Type 1 paradox in that a Type 2 paradoxical statement is false. The presentation is flawed and misleading. Once the fallacious step is noted, the paradox is exposed. So, the Type 2 paradox asks of the reader, "What's wrong with this picture? Where is the error?"

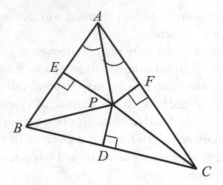

Figure 2.3. Are all triangles isosceles?

As an example of a Type 2 paradox, we consider a geometric *proof* that all triangles are isosceles (have at least two equal sides). The reader is encouraged to follow the argument presented and find the fallacious step.

The apparent proof that all triangles are isosceles begins by drawing the general triangle *ABC* (Figure 2.3). Then proceed as indicated.

Step 1.   Draw the line bisecting angle *A*.

Step 2.   Draw the line bisecting *BC* that is perpendicular to *BC*. Assume it intersects *BC* at *D*.

Step 3.   If the lines from the above two steps coincide then triangle *ABC* is isosceles. So, assume this is not the case (the lines are not parallel) and they intersect at some point *P*.

Step 4.   Draw the lines from *P* to *E* and *P* to *F* that are perpendicular to *AB* and *AC*, respectively.

Step 5.   The two triangles *APE* and *APF* are congruent because they are both right triangles sharing a common hypotenuse (*AP*) and having equal corresponding angles.

Step 6.   Since corresponding parts of congruent triangles are equal, *PE* = *PF*.

Step 7.   Note that as constructed, triangles *BDP* and *PDC* are right triangles.

Step 8.   The two right triangles have corresponding legs equal. This is because they share a common leg *DP* and *BC* is bisected by *D*. Consequently, the two triangles are congruent.

Step 9.   Since corresponding parts of congruent triangles are equal, *PB* = *PC*.

Step 10.   The right triangles *BEP* and *PFC* are now shown to be congruent as *PB* = *PC* (corresponding parts of congruent triangles) and *PE* = *PF* (corresponding parts of right triangles).

Step 11.   Since *BE* = *CF* and *EA* = *FA* (corresponding parts) we get *BA* = *CA* and the original triangle has now been shown to be isosceles.

A similar fallacious argument can be used to show *BC* = *BA*, in which case triangle *ABC* will have been shown to be equilateral, generalizing to the equally absurd conclusion that all triangles are equilateral. Readers unable to spot the fallacious step can see the discussion given in this chapter's appendix.

In *Mathematical Puzzles and Diversions*, Martin Gardner gives [Gardner 59, pp. 146–147] the following paradox, unique in that it involves two paradoxes, the first of Type 1 and the second of Type 2. A torus is a surface like that of a donut, or inner-tube. It is paradoxically (Type 1) true that by cutting a hole in the side of the torus, the surface can be turned inside out to form another torus, assuming the material is sufficiently elastic. (This can be verified by cutting the toes off of two socks and sewing them together to form the torus. Cut a two inch diameter hole in the side and the entire surface can be turned inside out.)

Now, start with a torus with one ring painted on the outside (solid line) and a second ring painted on the inside (broken line), as shown in the top illustration in Figure 2.4. The rings are linked.

By passing the torus through the small hole in its side, the torus can be turned inside out with the outside ring moving to the inside and the inside ring moving to the outside. But now it appears, as the bottom illustration in Figure 2.4 shows, that the two rings have become disengaged. Have the rings magically passed through each other? This impossibility suggests a Type 2 paradox. Where's the error? An explanation is given in this chapter's appendix.

But what if a statement, or argument, leads to logical inconsistencies which are not resolvable? This Type 3 paradox is an *antinomy*, defined by Webster as "a contradiction between two equally valid principles or between inferences correctly drawn from such principles." The word is used more in philosophical or religious contexts but can be applied

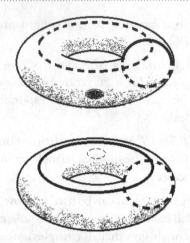

Figure 2.4. Inside out torus paradox.
(Reprinted with the permission of Martin Gardner.)

to mathematical inconsistencies as well. An antinomy is considered an extreme form of paradox, perhaps having no universally accepted resolution.

Russell's Paradox, or *Russell's Antinomy*, was discovered by the British mathematician Bertrand Russell and presented in 1901. Formally, it considers the set of all sets which are not members of themselves. Is this set a member of itself? Either answer leads to an immediate contradiction.

An alternative version of the antinomy, known as the *Barber Paradox*, was introduced by Russell in 1918. "If the town barber shaves the men, and only those men, who do not shave themselves, then who shaves the barber?" As before, whether we assume the barber shaves himself or does not shave himself, we are led to an immediate contradiction.

There is no simple resolution to such self-referential statements. Alfred Tarski, co-author of the Banach-Tarski Theorem, proposed the concept of *metalanguages* to deal with such problems, creating a hierarchy of metalanguages, each dealing with the truth or falsity of the languages below it.

Russell proposed an equivalent approach, which he called *theory of types*. Essentially, it suggests that statements like "a set is a member of itself" or "a set is not a member of itself" are meaningless. The theory denies the existence of self-contradictory sets.

There are many equivalent variations to these self-referential contradictions. Among them are the following:

"Divide all adjectives into two groups: those which describe themselves and those which do not. Into which group should the word 'indescribable' go?"

"I am a liar."
"Every general statement is false."

Jigsaw Fallacies

There are thousands of geometric paradoxes to be found in books and periodicals of recreational mathematics. This chapter concludes with a collection of jigsaw fallacies (Type 2 paradoxes) similar in appearance to the Banach-Tarski Paradox. There is, however, one very significant difference. The following examples are deceptions. They are misleading, with the sleight being difficult to detect. In comparison, the paradoxes given in Chapters 4 and 5, including the Banach-Tarski Paradox, are Type 1 paradoxes, and, therefore, true in actual fact.

One collection of jigsaw fallacies could properly be called *Loyd Vanishes* in that they were inspired by the great American puzzlist Sam Loyd (1841–1911). Loyd created thousands of puzzles, many of which were published after his death, by his son, Samuel Loyd, Jr. In 1896, he patented his *Get Off the Earth* puzzle, arguably his most famous invention. A rotating disc, representing the earth, is attached to a rectangular backboard. Thirteen Chinese warriors are painted around the circumference so that part of each warrior is on the earth (disc) and part is on the backboard. When the earth is positioned with the arrow

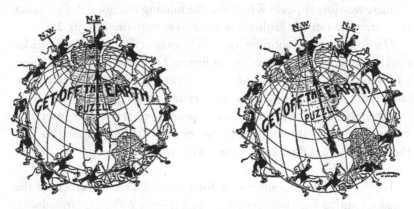

Figure 2.5. Sam Loyd's Get Off the Earth puzzle [Gardner 56, Figures 64 and 64a]. (Reprinted with the permission of Dover Publications, Inc.)

57

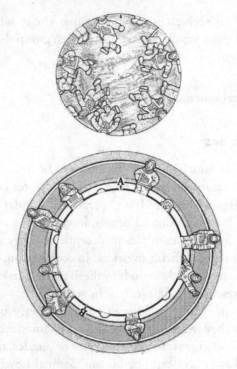

Figure 2.6. *The Vanishing Astronaut.*
(Reprinted by permission of the AIMS Education Foundation from *Puzzle Play*
© 2001 AIMS Education Foundation. All rights reserved.)

pointing to the NE pole, thirteen Chinese warriors can be counted. When the disc is rotated so the arrow points to the NW pole, only twelve Chinese warriors appear. Where did the missing warrior go? Reversing the process causes the vanished warrior to materialize (Figure 2.5).

The deception is beautifully subtle. To discover the concept, the reader could photocopy the illustrations in Figure 2.6, representing a variation which could be called *The Vanishing Astronaut* [Youngs 02]. With scissors, cut out the top disc and place it in the center of the ring below.

When the arrow of the inner disc is positioned at *A* of the outer rim, fifteen astronauts are visible. When the inner disc is rotated so that the arrow is at the *B* position, only fourteen astronauts appear? Which one vanished?

The W. A. Elliott Company of Toronto sold a clever version of the paradox entitled *The Vanishing Leprechaun* (Figure 2.7). It is formed with three rectangular pieces of cardboard assembled as one large rectangle.

Fifteen leprechauns

Fourteen leprechauns

Figure 2.7. *The Vanishing Leprechaun.*

Two small rectangles are on top and the large rectangle is on the bottom. It is a linear array of leprechauns with one part of each leprechaun on the upper half and one part on the lower half. When the two small upper rectangles are interchanged, a leprechaun vanishes.

If the reader has been unable to detect the deception in the *Loyd Vanishes* presented to this point, then the following materialization, used by some counterfeiters, should clarify matters.

Begin with nine dollar bills as shown in Step 1 of Figure 2.8. Next, cut each bill vertically as shown in Step 2. In Step 3, transfer the right portion of each bill down one level to the bill underneath. Finally, glue the pieces together forming ten bills, as shown in Step 4.

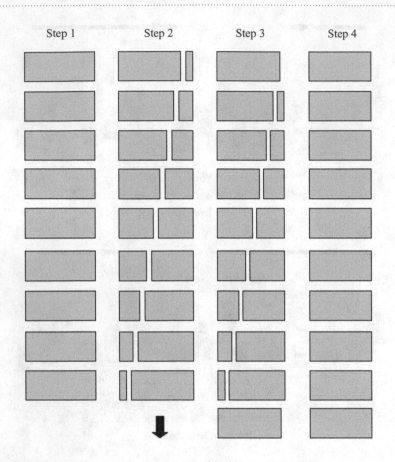

Figure 2.8. Nine bills become ten.

It should now become clear what is really happening with these deceptions. By shifting each right portion down to the next bill, the lengths of each of the new bills is 9/10 of the original length. The 10% decrease in length is difficult to perceive, unless the altered bills are placed side-by-side with a correct bill. Effectively, a new bill has been created by each of the original nine bills *donating* 10% of their length.

This is much the same as stretching four servings of peas to five by shaving off a bit from each of the four servings to create the fifth serving.

A close inspection of the *Get Off The Earth* puzzle will show that the height of each Chinese warrior has slightly increased after the vanish occurs. Similarly, the astronauts and leprechauns also increase in height, after the vanish. Despite the simplicity of the explanation, the deceptions have to be cleverly illustrated to disguise the secret.

The Loyd vanish/materialization has been exploited by counterfeiters with limited success. As would be expected, it is a U.S. federal offense to cut up bills in this manner, violating Chapter 25, Title 18 of U.S. Code Section 484. The November 29, 1968 edition of the *Chicago Daily News* reported that Reuben Silver of London had been convicted of doing this sort of thing with British 5 pound notes. "The court said the system was so good that details can't be made public" [Gardner 76, p. 34]. Note that with respect to U.S. currency, altered bills are easy to detect by checking the two serial numbers printed on each side of the bill. Since altered bills are composed of two pieces from separate bills, the serial numbers of the altered bills will not match.

Another scheme, *clipping*, involves the shaving of metal from coins containing the precious metals gold and silver. The Greeks and Romans put beaded borders on some coins to discourage clippers, with little success. Clipped coinage remained a nuisance until the mid-seventeenth century when, with improved technology, machines were introduced in England that put an intricate design on the coin's edge. A 1662 English crown shows on its edge, "DECVS ET TVTAMEN" ("An Ornament and a Protection"). The new process made clipping almost impossible and the use of the machines spread throughout Europe.

Today U.S. coins have small ridges, known as *reeds*, on the edges of dimes, quarters, half-dollars, and dollars. Despite the fact these coins no longer have a precious metal content, the reeding process continues for cosmetic reasons as well as being an identifier for the visually impaired.

And then there is the story of the computer programmer who, as a bank employee, rounded customers' interest payments down to the nearest cent and scraped off the excess into a special account—the programmer's. The swindle succeeded for a brief period of time because the amount of money improperly removed from each account, small when compared to the value of the account, went unnoticed by the account's owner.

The next group of jigsaw fallacies involves a sleight of a different sort. Each may be presented as either an appearance or disappearance (vanish) and the deception can be subtle if the figures are carefully illustrated.

The first shows the decomposition of an 8×8 square, or chessboard, into four pieces, with dimensions as shown in Figure 2.9. The original 8×8 square has an area of 64 square units.

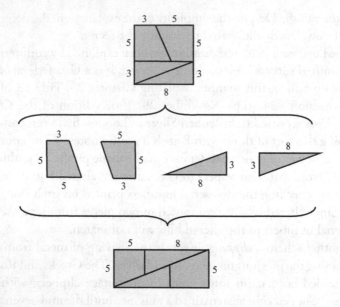

Figure 2.9. Paradoxical chessboard dissection.

The pieces are then reassembled to form a 5 × 13 rectangle, now of area 65. So, by decomposition and translation, there appears to be an unexplainable gain of one square unit. If the process is presented in reverse, there would be a vanish of one square unit (Figure 2.9).

Does the reader see the deception?

In this example, the paradox occurs as a result of a misleading illustration. The four pieces will fit into a 5 × 13 rectangle, but not quite as shown in Figure 2.9.

The long diagonal from the lower left to the upper right of the 5 × 13 picture does not really exist. The pieces come together in such a way that a space, in the form of a thin parallelogram, is created. Figure 2.10 shows the gap, slightly exaggerated.

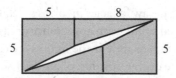

Figure 2.10. Mind the gap!

8 x 8 = 64   Cut and shift   9 x 7 = 63

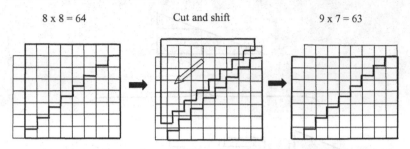

Figure 2.11. Another chessboard dissection.

The area of this thin parallelogram gap is exactly one square unit, which accounts for the one unit increase in overall area of the rectangle. As Londoners are warned when they step from the platform to the train, "Mind the gap!"

Speaking of chessboard dissections, the following jigsaw fallacy is found at Alex Bogomolny's Java interactive website *Cut The Knot* (http://www.cut-the-knot.org). Anyone with an interest in recreational mathematics should pay the site a visit.

An 8 × 8 chessboard is split, stepwise, along the diagonal as shown in Figure 2.11. When the two pieces are reassembled, a 9 × 7 rectangle is formed. The rectangular area has gone from 64 to 63 and there has been a paradoxical loss of one square unit.

Where did the missing square go?

Try to solve this visually, without using a ruler and without redrawing the figure.

Triangles, as well as rectangles, can be dissected and reassembled paradoxically. The dissected triangle in Figure 2.12 is sometimes called a *Curry Triangle*, named after mathematician and magician Paul Curry. Martin Gardner shows many variations of this theme in *Mathematics, Magic and Mystery* [Gardner 56, pp. 145–150].

The original right triangle, having a base leg of length 13 and a vertical leg of length 5 is dissected into four parts as shown in Figure 2.12. When the parts are reassembled, they neatly fit back into an identical triangle with one square unit to spare. So, once again, there is a paradoxical loss of one square unit. Careful inspection by the reader should reveal the deception.

Paul Curry combined the vanish concepts discussed to create the "Disappearing Rabbit Paradox," shown in Figure 2.13.

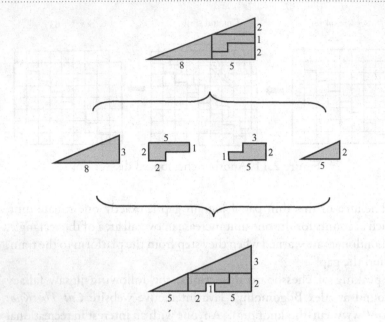

Figure 2.12. Curry triangle.

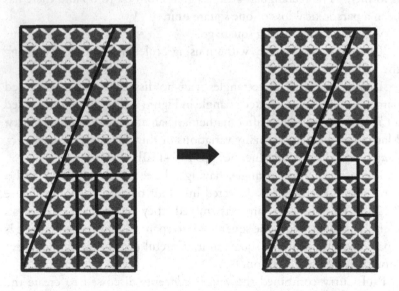

Figure 2.13. The Disappearing Rabbit Paradox.
(Courtesy of Myles Curry.)

The 6 × 13 rectangle on the left contains 78 squares, each containing a rabbit. The large rectangle is then dissected into five pieces and reassembled, forming another 6 × 13 rectangle. But now it appears as if one of the rabbits has gone missing. Note that each of the five subsections on the left is paired, exactly, with one of the subsections on the right.

How do we account for the missing rabbit?

Recreations are entertaining but the stage is now set to explore the famous theorem of Banach and Tarski. Chapter 3 is a *back to school* review of the mathematics needed to fully appreciate the theorem and its proof. Topics reviewed include elementary set theory and plane geometry. In addition, some new concepts may be introduced, depending on the reader's mathematics background.

Then it is on to Chapter 4 where the Baby BTs are presented. These are mathematically valid dissections and reassemblies resulting in an apparent gain. Keep in mind that the jigsaw fallacies just presented in Chapter 2 are just that—fallacies. The Baby BTs, on the other hand, are not fallacies. The constructions are simple, the results are Type 1 paradoxical, and the conclusions are valid.

We take direct aim at the Banach-Tarski Theorem in Chapter 5 —the focal chapter of this book. As is the case with the Baby BTs, the consequences of the theorem are remarkable, yet valid. Such results force mathematicians to reconsider the nature of mathematics and how it relates to the physical world.

Appendix

Simpson's Paradox is resolved by considering the factors which contribute to the overall success rate being studied. In the example given, medication type (red pills or yellow pills) is one factor and sex of the patient is another. Note that the male population survives at a significantly higher rate than the female population, regardless of the medication.

The red pills are, in fact, more effective than the yellow pills and should be the treatment of choice by patient and physician. The pooling of the data masks this, suggesting the yellow pills are more effective, as a result of the fact that the yellow pills were given, almost exclusively, to male patients, who do much better, across the board, than the female patients. The pooled data paradoxically exhibits success of the yellow pills, which in reality is survival of males. It would be incorrect to attribute the survival to the yellow pills. Caution must be used when data is pooled to avoid the Simpson phenomenon.

Real world occurrences of Simpson's Paradox have been documented. The following three are noted by C. H. Wagner [Wagner 82, pp. 46–48]:

1. The subscription renewal rate for *American History Illustrated* increased from January, 1979 to February, 1979, despite the fact the rate decreased for each category of subscriber.

2. The overall federal tax rate increased from 1974 to 1978 yet decreased for each income bracket.

3. The overall tuberculosis death rate in 1910 was greater in Richmond, Virginia than New York City; however, it was less for whites and non-whites during the same period.

A classic occurrence of Simpson's Paradox with respect to sex bias has been well documented [Bickel, Hammel, and O'Conell 75]. For the fall quarter of 1973, the University of California at Berkeley accepted 44% of male applicants to its graduate program compared to 35% of female applicants. The selections were made from a pool of approximately 15,000 applicants, suggesting sex bias in favor of male applicants with the possibility of sex discrimination. Yet, when broken down by department, there was no bias favoring men. Interestingly, there was a slight bias favoring women. Why did pooling cause the reversal?

Men tended to apply to those departments that accepted a relatively high percent (60% to 80% of all applicants); whereas women tended to apply to more competitive departments with a lower acceptance rate for both males and females. So, when the results were pooled, there was the suggestion of sex discrimination. As always, caution must be used when interpreting pooled data.

The Inside Out Torus Paradox, being Type 2, is clarified by noting the error in its presentation. It is correct that the torus can be turned inside out to form another torus.

It is also correct that the outside ring will pass to the inside and the inside ring will pass to the outside. The bottom illustration in Figure 2.4, however, is incorrect. The rings, of course, can not pass through each

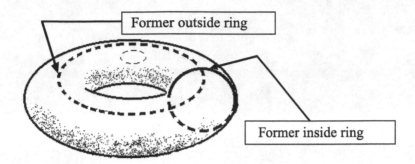

Figure 2.14. Correct view of torus turned inside out.
(Reprinted with the permission of Martin Gardner.)

other, and the correct orientation, after turning the torus inside out, is shown in Figure 2.14. The rings remain linked.

Poorly drawn figures are the culprit behind many Type 2 paradoxes, including the so-called proof that all triangles are isosceles. The deception occurs in the third step of the proof given. Point P, the intersection of two non-parallel straight lines, was shown as being inside of triangle ABC. In actuality, it is outside of triangle ABC as shown in Figure 2.15.

With point P in its correct position, exterior to triangle ABC as shown, the proof that the triangle is isosceles breaks down at Step 11, the last step of the fallacious proof.

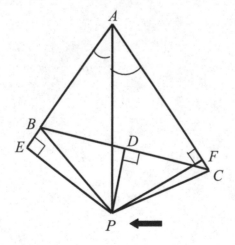

Figure 2.15. Point P is outside triangle ABC.

3 Preliminaries

Math class is tough!
—Barbie's 1992 voice chip by Mattell™

The length of the proof of the Banach-Tarski Theorem depends on the sophistication of the mathematics used. The proof can be presented on a single page if it assumed the reader is familiar with basic set theory, equidecomposability, and the Hausdorff Paradox. Here we assume no such thing. So the proof given in Chapter 5 will be a bit lengthy, but elementary. It's much like building a house—the more sophisticated the equipment, tools and labor, the quicker the job can be accomplished. The same house could be built with nothing more than crude hand tools, but it would take significantly longer to complete the job.

Despite our elementary approach, we must review the following four topics before beginning the proof:

1. Set theory
2. Isometries
3. Scissors congruence
4. Equidecomposability

Much of what we review is quite simple and can be taught to grade school students. Set theory, for example, was introduced to U.S. elementary school curriculum in the 1960s as part of the so-called *new math*. There was incentive to beef up the science and mathematics curriculum in the United States following the Soviet Union's launch of the Sputnik satellite on October 4, 1957. A decade later, much of the previously introduced set theory was removed from the curriculum as it was felt students were not getting enough drill and practice in basic computation; set theory could be taught at a later time.

The reader must not associate complex notation with complex mathematics. Much of this chapter is intuitive and every attempt is made to keep the notation elementary.

Set Theory

Definition of Set

A set is a collection of objects. Each object in the set is called an element, or member of the set.

Examples of sets would include the set of all living U.S. citizens, the set of all real numbers greater than five, and the set of all points in three-dimensional space which are within one unit of a given point. The last example is a *point set* and, in this case, represents a solid sphere (or ball) of radius one. The proof of the Banach-Tarski Theorem involves point sets and sets of rotations. We will discuss rotations in the isometries section of this chapter.

There are basically three ways to define a set. The most obvious way is to describe it, as was done for the previous examples. For sets consisting of mathematical objects (numbers, variables, points, etc.) we can be more precise. We write $A = \{a, b, c\}$ to indicate the set A is to consist of the three elements $a, b,$ and c. This is the *list* or *roster* method of set description. We could specify the set of counting numbers from one to ten as $B = \{1, 2, 3,..., 10\}$. The list method works well for small sets where each element can be precisely specified.

A third option, *set builder notation*, is a formal way of describing sets not easily listed. The set of all points (x, y, z) in three-dimensional space satisfying the equation $x^2 + y^2 + z^2 = 1$ could be specified as $\{(x, y, z) : x^2 + y^2 + z^2 = 1\}$. This three-dimensional point set is the sphere of radius one centered at (0, 0, 0).

The symbol $\in$ denotes set membership. So we write $1 \in \{1, 2, 3\}$ to indicate that the number 1 is an element of the set $\{1, 2, 3\}$ Similarly, we write $4 \notin \{1, 2, 3\}$. The *cardinality* of a set is simply the number of elements in the set. Cardinality is a measure of set content, telling us "how many". For example, if $A = \{1, 5, 7, 9, 20\}$ then the cardinality of A is five and we write $n(A) = 5$ to denote this fact. Recall from Chapter 1 that the cardinality of the set of counting numbers is $\aleph_0$ and the cardinality of the continuum is c. More will be said of infinite cardinalities later in this chapter.

In any given application of set theory, there are two special sets of extreme cardinality. The first is the *null*, or *empty set*, which has no members. It is specified by the symbol $\emptyset$ or $\{\,\}$. Note $n(\emptyset) = 0$. The other is the *universal set U*, which by definition is the set of all elements under consideration for the application. Think of the universal set as the environment in which the sets of the application reside. For example, when doing a statistical study of U.S. voter preferences, the universal set U might be the set of all U.S. registered voters. When working with point sets in three-dimensional space, the universal set would be the set of all points in three-dimensional space.

Definition of Subset

Set A is a subset of set B if and only if each element of A is also an element of B. We write $A \subseteq B$ to denote the fact that A is a subset of B.

So, the set of rational numbers is a subset of the real numbers and the set of positive even numbers is a subset of the set of counting numbers. It follows that if A is not a subset of B then A must contain at least one element which is not an element of B. It is for this reason that the null set, $\emptyset$, is considered to be a subset of any given set. For if it were not, we should be able to find an element in $\emptyset$ which is not in the given set. Clearly this is impossible.

Examples:

$\{1, 2, 3\} \subseteq \{1, 2, 3, 4, 5\}$

$\{x : 0 \le x \le 1\} \subseteq \{x : 0 \le x < \infty\}$

$\emptyset \subseteq \{1, 2, 3\}$

Definition of Equality of Sets

Two sets, A and B, are equal if they contain the exact same elements. Formally, $A = B$ if $A \subseteq B$ and $B \subseteq A$.

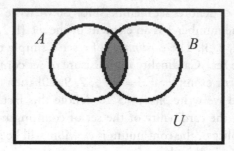

Figure 3.1. The shaded area is $A \cap B$.

The definition is intuitive and reminds one of the equality of real numbers. (Real numbers a and b are equal if $a \le b$ and $b \le a$)

From a given set or collection of sets, we can define other sets using standard set operations. Just as the set of real numbers has the basic operations of addition, subtraction, multiplication, and division, set theory has its basic operations. Some definitions follow.

Definition of Set Intersection

The intersection of two sets A and B, denoted by $A \cap B$,
is the set of all elements belonging to both A and B. Formally,
$$A \cap B = \{x : x \in A \text{ and } x \in B\}.$$

Examples:

$\{1, 2, 3\} \cap \{1, 3, 5\} = \{1, 3\}$

$\{1, 3, 5\} \cap \{2, 4, 6\} = \emptyset$

$\{1, 2, 3\} \cap \emptyset = \emptyset$

The Venn diagram in Figure 3.1 illustrates the concept of intersection for point sets in the plane.

Definition of Set Union

The union of two sets A and B, denoted by $A \cup B$, is
the set of all elements belonging to either A or B. Formally,
$$A \cup B = \{x : x \in A \text{ or } x \in B\}.$$

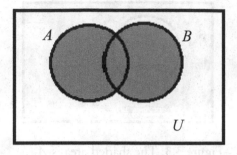

Figure 3.2. The shaded area is $A \cup B$.

Examples:

$\{1, 2, 3\} \cup \{1, 3, 5\} = \{1, 2, 3, 5\}$

$\{1, 3, 5\} \cup \{2, 4, 6\} = \{1, 2, 3, 4, 5, 6\}$

$\{1, 2, 3\} \cup \emptyset = \{1, 2, 3\}$

The Venn diagram in Figure 3.2 illustrates the concept of union for point sets in the plane.

Definition of the Complement of a Set

The complement of a set A, denoted by A^C, is the set of elements in U that are not in A. Formally, $A^C = \{x : x \in U, x \notin A\}$.

Note that the complement of a set cannot be specified unless we have a clear definition of the universal set.

Examples:

Let $U = \{1, 2, 3, \ldots, 10\}$, $A = \{1, 2, 3\}$, and $B = \{2, 4, 6, 8, 10\}$. Then

$A^C = \{4, 5, 6, 7, 8, 9, 10\}$

$(A \cup B)^C = \{1, 2, 3, 4, 6, 8, 10\}^C = \{5, 7, 9\}$

$U^C = \emptyset$

$\emptyset^C = U$

The Venn diagram in Figure 3.3 illustrates the concept of the complement of a point set in the plane.

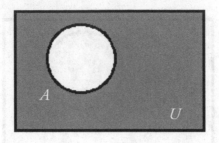

Figure 3.3. The shaded area is A^C.

<u>Definition of the Difference of Sets</u>

Let A and B be two sets. The difference A − B is the set consisting of all elements in A that are not in B. Formally,
$$A - B = \{x : x \in A \text{ and } x \notin B\} = A \cap B^C.$$

Examples:

Let $U = \{1, 2, 3, \ldots, 10\}$, $A = \{1, 2, 3\}$, and $B = \{2, 4, 6, 8, 10\}$

Then

$A - B = \{1, 3\}$

$B - A = \{4, 6, 8, 10\}$

The Venn diagram in Figure 3.4 illustrates the concept of set difference for point sets in the plane.

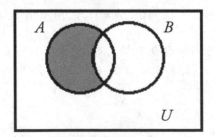

Figure 3.4. The shaded area is $A - B$.

Definition of Power Set

*For any set A, the power set of A, denoted as $\mathscr{P}(A)$,
is the set of all subsets of A.*

Example:

Let A = {1, 2, 3}. Then $\mathscr{P}(A)$ = {∅,{1},{2},{3},{1, 2},{1, 3}, {2, 3}, {1, 2, 3}}.
There are eight subsets shown as members of the power set. We must
be careful to include ∅ and {1, 2, 3} as these are indeed subsets.

For any finite set A it can be shown that the cardinality of the power
set of A is $2^{n(A)}$. That is $n[\mathscr{P}(A)] = 2^{n(A)}$. The proof is by mathematical
induction and we omit it here. For the previous example where
$A = \{1, 2, 3\}$ the cardinality of A is 3 and the cardinality of $\mathscr{P}(A)$ is
$2^3 = 8$.

We have seen nothing remarkable or paradoxical at this point in our
discussion of set theory because finite sets are easily comprehended.
Strange and paradoxical results arise when we consider sets of infinite
cardinality. We briefly discussed two such sets in Chapter 1—the
counting numbers $\{1, 2, 3, ..\}$ and the set of all real numbers which we
will denote as $\mathbb{R}$. Recall that the cardinality of $\{1, 2, 3, . . .\}$ is denoted
by aleph naught— $\aleph_0$, which is also the cardinality of the set of rational
numbers. The cardinality of the set of real numbers $\mathbb{R}$ is denoted by
c. This is the cardinality of the continuum, which also represents the
number of points on the real number line between zero and one. We
established that $\aleph_0 < c$ by giving Cantor's diagonalization proof that
there are more real numbers than counting numbers. We are motivated
to write $c = 2^{\aleph_0}$ by each of the following arguments:

1. Any real number between zero and one can be written uniquely
    in binary (base two) notation as $.b_1b_2b_3...b_n...$ where each b_i
    is either zero or one. Each representation is unique if we do not
    allow a representation to terminate. Terminating representations
    can be avoided by replacing any terminating representation with
    its equivalent nonterminating representation ending in infinitely
    many 1s. For example, we write .011111 . . . instead of .1 and
    .1011111 . . . instead of .11. Since there are two possibilities for each
    digit (zero or one) the total number of possible real numbers between
    zero and one can be thought of as the product of a countable number
    of 2s. That is, the cardinality of the continuum is $c = 2^{\aleph_0}$.

2. Being more precise, we show there is a one-to-one correspondence between the set of real numbers between zero and one (cardinality c) and the power set of the counting numbers (cardinality $2^{\aleph_0}$). We do this in two steps. First, we show there is a one-to-one function, or mapping (technically termed an *injection),* from the set of real numbers between zero and one into the power set of the counting numbers. This establishes $c \leq 2^{\aleph_0}$. Then we show there is a one-to-one function from the power set of the counting numbers into the set of real numbers between zero and one. This establishes $2^{\aleph_0} \leq c$. Since $2^{\aleph_0} \leq c$ and $c \leq 2^{\aleph_0}$ we intuitively conclude $c = 2^{\aleph_0}$. (A formal proof of this fact requires the Schröder-Bernstein Theorem. We prove a version of this theorem in Chapter 5.)

   To accomplish the first step of this argument, identify each real number between zero and one with its non-terminating binary representation $.b_1b_2b_3\ldots b_n\ldots$. To this real number we associate a subset of the counting numbers which includes the counting number i if and only if $b_i = 1$. This establishes a one-to-one function of the real numbers between zero and one into the power set of the counting numbers. That is, $c \leq 2^{\aleph_0}$.

   To establish the reverse inequality, we must find a one-to-one function from the power set of the counting numbers into the set of real numbers between zero and one. To each subset of the counting numbers define a base ten decimal fraction $.d_1d_2d_3\ldots d_n\ldots$ where $d_i = 1$ if and only if i is in the subset of counting numbers; otherwise set $d_i = 0$. This establishes a one-to-one function of the power set of counting numbers into the set of real numbers between zero and one. That is $2^{\aleph_0} \leq c$. Since $2^{\aleph_0} \leq c$ and $c \leq 2^{\aleph_0}$, we claim $c = 2^{\aleph_0}$.

Are there other transfinite numbers? Is there a largest transfinite number? The following theorem establishes a never ending ascending sequence of transfinite cardinals.

Cantor's Theorem

For any set A, $n(A) < n[\mathscr{P}(A)]$.

Proof:

   Clearly $\mathscr{P}(A)$ must have at least as many elements as A because to each $a \in A$ we can associate $\{a\} \in \mathscr{P}(A)$. It remains to be shown that

$n(A) \neq n[\mathscr{P}(A)]$. We do this by first by assuming a one-to-one correspondence exists and then showing this leads to a contradiction.

If we assume there is a one-to-one correspondence between A and $\mathscr{P}(A)$, then there exists a one-to-one and *onto* function (mapping) f from A to $\mathscr{P}(A)$. (A mapping from one set to another is *onto* if each element of the second set is mapped to by at least one element of the first set. If a mapping is both *one-to-one* and *onto*, then each element of the second set is mapped to by a *unique* element of the first set. In such a case there is a one-to-one correspondence between the sets. A further disucssion of these terms is given in Chapter 4.) For arbitrary $a \in A$ it may (or may not) be that $a \in f(a)$. Let $B = \{x : x \in A, x \notin f(x)\}$. Note $B \subseteq A$ hence $B \in \mathscr{P}(A)$. By assumption f is onto so there must exist $b \in A$ such that $f(b) = B$.

Is $b \in f(b)$? Either answer leads to a contradiction. If $b \in f(b)$ then $b \in B$ and $b \notin f(b)$. If $b \notin f(b)$ then $b \in B$ and $b \in f(b)$. The only way out of this is to assume f cannot exist which implies $n(A) < n[\mathscr{P}(A)]$.

So, if we start with the counting numbers and form the power set of the counting numbers and then the power set of the power set of the counting numbers, etc., we get the following ascending transfinite cardinalities: $\aleph_0 < 2^{\aleph_0} < 2^{2^{\aleph_0}} < 2^{2^{2^{\aleph_0}}}, \ldots$. Hermann Weyl critically referred [Stewart 96, p. 67] to these cardinalities as "fog on fog."

Are there additional transfinite cardinal numbers between those of the above sequence?

We'll never know! At least we'll never know within the Zermelo-Fränkel axiomatic system. Cantor proposed the existence of a hierarchy of infinite cardinalities beginning with the smallest infinity—$\aleph_0$. Using the aleph notation, Cantor proposed infinitely many transfinite cardinalities—$\aleph_0 < \aleph_1 < \aleph_2 < \ldots < \aleph_n < \aleph_{n+1} < \ldots$ where $\aleph_{n+1}$ was the next largest transfinite cardinality after $\aleph_n$, for $n = 0, 1, 2, \ldots$. The question remains as to whether c is the next largest transfinite cardinality after $\aleph_0$. That is, does $\aleph_1 = c$?

Cantor believed so and the reader may recall this from Chapter 1 as the famous Continuum Hypothesis. Cantor was obsessed with proving it and was unable to do so. It took the efforts of Kurt Gödel and Paul Cohen to establish it as undecidable. A more generalized version of the

Continuum Hypothesis was proposed by Felix Hausdorff in 1908. It links all transfinite cardinalities by the equation $\aleph_{n+1} = 2^{\aleph_n}$. It too was established as being undecidable by Gödel and Cohen.

As presented in Chapter 1, there are several well known paradoxes associated with Zermelo-Fränkel set theory. The three presented next can be resolved, to some extent, by techniques beyond the scope of this presentation. They all possess a self-referential *set of all sets* theme, as discussed in Chapter 1. None of these three paradoxes involve the Axiom of Choice. Paradoxes arising from this axiom will be given in Chapters 4 and 5.

Historically, the first paradox was discovered by Burali-Forti in 1897. To be precise, it is a statement about *ordinal,* as opposed to *cardinal* numbers. The cardinal number 3 is a measure of totality. It represents a measure of *content* or *manyness*. The ordinal number 3 is a measure of *rank* or *position*. If you are 3rd in line, we are using the number 3 as an ordinal number. If the bucket contains 3 apples, we are using the number 3 as a cardinal number.

Burali-Forti Paradox

An ordinal number, representing rank, can be defined set theoretically as the set of all its predecessors.

Symbol	Set Definition
0	{}
1	{{}}
2	{ {}, {{}} }
.	
.	

In general n can be defined as $\{0, 1, 2, 3, \ldots, n-1\}$.

The set of all ordinal numbers would then define an ordinal number greater than any ordinal number in the set. Therefore the set of all ordinal numbers do not form a set. This is self-contradictory.

Cantor's Paradox is the cardinal number analogy of the Burali-Forti Paradox.

Cantor's Paradox

Let S denote the set of all sets. Every subset of S, being a set, must also be a member of S. It follows that the power set of S is a *subset of S. That is $\mathcal{P}(S) \subseteq S$. If so, then $n[\mathcal{P}(S)] \leq n(S)$ which clearly contradicts Cantor's Theorem.*

Russell's Paradox is a formal version of the Barber Paradox (antinomy) given in Chapter 2.

Russell's Paradox

Let Z be the set (or collection) of all sets which do not contain themselves as members. That is $Z = \{X : X \notin X\}$. Is Z a member of itself? If yes, then by definition of Z, Z is not a member of itself. If no, then by definition of Z, Z is a member of itself. Either way, we are led to a contradiction.

The Axiom of Choice, as discussed in Chapter 1, leads to additional paradoxes, most notable of which is the Banach-Tarski Paradox. Though discussed earlier, a formal version of the axiom is given, for the sake of completeness.

Axiom of Choice

For any collection C of nonempty sets, we can choose a member from each set in that collection. That is, there exists a choice function f defined on C such that for each set $S \in C$, $f(S) \in S$.

Isometries

Pick up a vase and place it elsewhere. Mathematically speaking, a set of points has been transformed (or mapped) to another set of points in such a way that the shape of the set being transformed (the vase) has been preserved. The distance between pairs of points on the vase remains the same when comparing the final position of the vase to the initial position of the vase. Such a distance preserving transformation is called an *isometry*. Think of an isometry as a rigid motion, or transformation, of a point set, without stretching, compressing, or some other distortion of the shape (Figure 3.5). A *preimage* gets mapped to an *image*; the two point sets are called *congruent*.

When, as dictated by the Banach-Tarski Theorem, we partition the ball and reassemble the pieces, we use isometries to move the individual pieces (point sets) of the ball. So we must be perfectly clear which types of transformations are isometries and which transformations are non-isometric. If non-isometric transformations were allowed, then creating a large volume from a smaller one could easily be done by stretching.

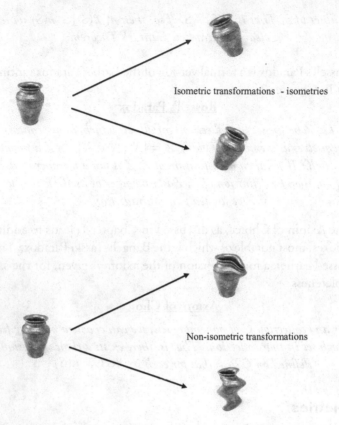

Figure 3.5. Continuous point set transformations.

There are four basic types of isometries and they are illustrated in Figure 3.6 for point sets in the plane. In each case, the preimage is congruent to the image.

The simplest isometry is the *translation*, also referred to as a *shift*. In the plane, the preimage point set is shifted, without rotation or distortion, to another position in the plane. A horizontal translation of a units is denoted as $(x, y) \rightarrow (x + a, y)$ where (x, y) is an arbitrary point of the preimage and $(x + a, y)$ is the point to which it is mapped. If $a > 0$, the shift is to the right and if $a < 0$ the shift is to the left. A vertical translation of b units can be similarly described as $(x, y) \rightarrow (x, y + b)$ The translation is up if $b > 0$ and down if $b < 0$. Similar notation is used for translating point sets in three-dimensional space.

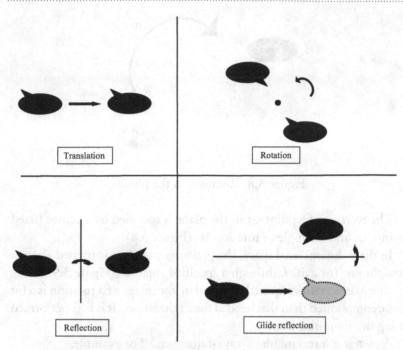

Figure 3.6. Fundamental isometries.

The two single translations described previously, if applied sequentially, would be equivalent to the single oblique translation $(x, y) \rightarrow (x + a, y + b)$. The translation of a translation is equivalent to a single translation. So in three dimensions if point (x, y, z) is first translated by T_1, and then T_2, we could describe the combined translation as $T_2[T_1(x, y, z)]$ or simply $T_2 T_1(x, y, z)$. The oblique translation $T_2 T_1$ is called the product of translations T_1 and T_2. Note the order of the subscripts. The translation $T_2 T_1$ represents T_1 followed by T_2 (Figure 3.7).

Figure 3.7. Product of T_1 and T_2.

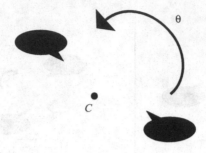

Figure 3.8. Rotation in the plane.

The *rotation* of a point set in the plane is specified by a center (fixed point) C and an angle of rotation θ (Figure 3.8).

In three-dimensional space, the preimage point set is rotated about a straight line (or axis) l, through a specified angle θ (Figure 3.9).

The mathematics required to calculate the image of a rotation is a bit more complicated than that needed for a translation. It is best performed using matrix multiplication.

A *matrix* is a rectangular array of numbers. For example,

$$A = \begin{bmatrix} 1 & 4 \\ 3 & 5 \\ 2 & 7 \end{bmatrix} \text{ is a } 3 \times 2 \text{ matrix,}$$

having three rows and two columns. A specific element of a matrix is denoted using lower case and double subscripts. So here $a_{12} = 4$ states

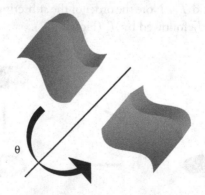

Figure 3.9. Rotation in space.

the number in the first row and second column is 4. Similarly, $a_{32} = 7$. A general $m \times n$ matrix could be written as

$$A = \begin{bmatrix} a_{11} & a_{12} & a_{13} & \cdot & \cdot & \cdot & a_{1n} \\ a_{21} & a_{22} & a_{23} & \cdot & \cdot & \cdot & a_{2n} \\ \cdot & & & & & & \\ \cdot & & & & & & \\ \cdot & & & & & & \\ a_{m1} & a_{m2} & a_{m3} & \cdot & \cdot & \cdot & a_{mn} \end{bmatrix}$$

If $m=n$, we refer to A as a square matrix. If $m=1$, A is called a row matrix (or vector) and if $n=1$, the matrix is a column matrix (or vector).

There are various operations associated with matrices but we need only be concerned with matrix multiplication, as it is the operation necessary for rotation of point sets. Let A be an $m \times n$ matrix and let B be an $n \times q$ matrix. To form the matrix product AB it will be required that the columns of A equal the rows of B. Let $[a_1\ a_2 \ .\ .\ .\ a_n]$ be a typical row of A and let

$$\begin{bmatrix} b_1 \\ b_2 \\ b_3 \\ \cdot\ \cdot\ \cdot \\ b_n \end{bmatrix}$$

be a typical column of B. The *inner product* of this row and column is given by $a_1 b_1 + a_2 b_2 + a_3 b_3 + \ldots + a_n b_n$. We now define the product AB of two matrices.

The Product of Two Matrices

Let A be an $m \times n$ matrix and B be an $n \times q$ matrix. Then the product matrix AB is the $m \times q$ matrix whose entry in the ith row and jth column is the inner product of the ith row of A and the jth column of B.

Example:

$$\begin{bmatrix} 5 & 3 \\ -3 & 2 \end{bmatrix} \begin{bmatrix} 4 & 0 \\ 3 & 1 \end{bmatrix} = \begin{bmatrix} 5 \cdot 4 + 3 \cdot 3 & 5 \cdot 0 + 3 \cdot 1 \\ -3 \cdot 4 + 2 \cdot 3 & -3 \cdot 0 + 2 \cdot 1 \end{bmatrix} = \begin{bmatrix} 29 & 3 \\ -6 & 2 \end{bmatrix}$$

$$\begin{bmatrix} 3 & 5 \\ 7 & 9 \end{bmatrix} \begin{bmatrix} 10 \\ 20 \end{bmatrix} = \begin{bmatrix} 3 \cdot 10 + 5 \cdot 20 \\ 7 \cdot 10 + 9 \cdot 20 \end{bmatrix} = \begin{bmatrix} 130 \\ 250 \end{bmatrix}$$

A rotation about the origin of a two-dimensional point set in the plane can be specified by a 2 × 2 matrix. To be precise, the rotation matrix is given by

$$R = \begin{bmatrix} \cos\theta & -\sin\theta \\ \sin\theta & \cos\theta \end{bmatrix}$$

where θ is the angle of rotation as measured in a counterclockwise direction. Readers not having a background in trigonometry need not be concerned as we need only to understand that rotations about $(0, 0)$ can be calculated by matrix multiplication.

For example, a rotation of 90° applied to the point $(1, 0)$ yields

$$\begin{bmatrix} 0 & -1 \\ 1 & 0 \end{bmatrix} \begin{bmatrix} 1 \\ 0 \end{bmatrix} = \begin{bmatrix} 0 \\ 1 \end{bmatrix}$$

representing an image point of $(0, 1)$. A rotation of $(1, 0)$ by 45° yields

$$\begin{bmatrix} \dfrac{\sqrt{2}}{2} & -\dfrac{\sqrt{2}}{2} \\ \dfrac{\sqrt{2}}{2} & \dfrac{\sqrt{2}}{2} \end{bmatrix} \begin{bmatrix} 1 \\ 0 \end{bmatrix} = \begin{bmatrix} \dfrac{\sqrt{2}}{2} \\ \dfrac{\sqrt{2}}{2} \end{bmatrix} \quad \text{representing the image point} \quad \left(\dfrac{\sqrt{2}}{2}, \dfrac{\sqrt{2}}{2} \right).$$

A 3 × 3 matrix is required to define the rotation of a three-dimensional point set about an axis through $(0, 0, 0)$. The trigonometry is technical and we omit it here. Two special rotations of this sort are needed for the proof of the Banach-Tarski Theorem given in Chapter 5. The first is

$$\tau = \begin{bmatrix} -\dfrac{1}{2} & -\dfrac{\sqrt{3}}{2} & 0 \\ \dfrac{\sqrt{3}}{2} & -\dfrac{1}{2} & 0 \\ 0 & 0 & 1 \end{bmatrix}$$

which represents a rotation of 120° about the z axis. The second is

$$\sigma = \begin{bmatrix} 0 & 0 & 1 \\ 0 & -1 & 0 \\ 1 & 0 & 0 \end{bmatrix}$$

representing a rotation of $180°$ about the straight line $z = x$ in the xz plane. As with translations, we can form products of rotations. The rotation τ followed by the rotation σ is equivalent to the single rotation $\sigma\tau$ which is computed as the matrix product of σ and τ. (Remember, τ applies first, then σ.)

The rotation matrix for $\sigma\tau$ is given by

$$\sigma\tau = \begin{bmatrix} 0 & 0 & 1 \\ 0 & -1 & 0 \\ 1 & 0 & 0 \end{bmatrix} \begin{bmatrix} -\dfrac{1}{2} & -\dfrac{\sqrt{3}}{2} & 0 \\ \dfrac{\sqrt{3}}{2} & -\dfrac{1}{2} & 0 \\ 0 & 0 & 1 \end{bmatrix} = \begin{bmatrix} 0 & 0 & 1 \\ -\dfrac{\sqrt{3}}{2} & \dfrac{1}{2} & 0 \\ -\dfrac{1}{2} & -\dfrac{\sqrt{3}}{2} & 0 \end{bmatrix}.$$

A single matrix specifies a rotation as long as the center is $(0,0)$ for point sets in the plane or the axis is a straight line through $(0,0,0)$ for three-dimensional point sets. Other rotations are more complicated to compute. One option would be to use a rotation matrix as we have done, then follow with the appropriate translation to correctly position the image.

Reflections and *glide reflections* are not used when the pieces of the Banach-Tarski decomposition are moved about and reassembled. We only use translations and rotations. However the reflection isometry has some properties worth mentioning. The reflection of a point set changes its orientation; e.g., the image of a right-hand glove is a left-hand glove. The movement associated with a reflection can't be executed within the original space. The set of points must pass through a space one dimension higher than that of the original space. For example, if we wish to reflect a two-dimensional point set about a straight line in its plane via a continuous rigid motion, it must be revolved out into three-dimensional space and then dropped back into the plane. A general two-dimensional point set can not be reflected by continuously sliding it around the plane. Similarly, if a three-dimensional point set is to be reflected through a plane (mirror), it cannot be done in one continuous rigid motion without passing through four-dimensional space.

As peculiar an isometry as the reflection appears, it should be considered the most fundamental of the isometries because any isometry can be defined in terms of reflections alone! In the plane, for example, a translation is the product (sequential application) of two reflections over parallel lines. A rotation is the product of two reflections over intersecting lines. The center of the rotation is the intersection of the lines and the angle of rotation is twice the measure of the non-obtuse angle between the lines. Despite the fact we could get by with reflections alone to describe isometries, it is better for our purposes to think in terms of translations and rotations. Reflections are more difficult to describe, mathematically.

The collection of all possible isometries associated with a set of points forms a mathematical structure known as a *group*. A group is a collection of objects along with a *binary operation* possessing the properties of *closure, associativity, identity*, and *inverse*. Each property is briefly described as it relates to an isometry group.

In this case, the objects are isometries and the binary operation is that of sequential application. When two isometries are combined sequentially and written as a product, we get the equivalent of a single isometry. In that combinations of isometries always yield isometries, the set of isometries is said to be *closed* under the binary operation of sequential application.

Sequential application is *associative*. That is, for any three isometries ρ_1, ρ_2, and ρ_3 it can be shown $(\rho_1\rho_2)(\rho_3) = \rho_1(\rho_2\rho_3)$.

There is a special isometry, I, called the *identity* isometry. For this isometry the image is identically equal to the preimage. It is analogous to multiplication by one or the addition of zero within the set of real numbers.

Every isometry ρ is associated with an *inverse* isometry, denoted by ρ^{-1}. Simply put, an isometry and its inverse reverse each other. One undoes the other. That is, $\rho\rho^{-1} = \rho^{-1}\rho = I$, the identity isometry. Note $(\rho_1\rho_2)^{-1} = \rho_2^{-1}\rho_1^{-1}$ since

$$(\rho_1\rho_2)(\rho_2^{-1}\rho_1^{-1}) = \rho_1(\rho_2\rho_2^{-1})\rho_1^{-1} = \rho_1 I\rho_1^{-1} = \rho_1\rho_1^{-1} = I.$$

Similarly, $(\rho_1\rho_2\cdots\rho_n)^{-1} = \rho_n^{-1}\rho_{n-1}^{-1}\cdots\rho_1^{-1}$.

In general, sequential application of isometries is not commutative. That is, in general, $\rho_1\rho_2 \neq \rho_2\rho_1$. Change the order of application and we get a different product isometry. We could easily show this mathematically, using the fact that matrix multiplication is not commutative. But, the illustration in Figure 3.10 should make it clear. A single die with opposite sides similarly shaded is rolled forward ¼ turn then right ¼ turn. Start over but reverse the order of rotations. The outcomes are different.

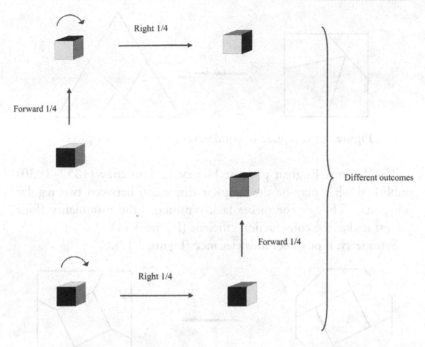

Right 1/4

Forward 1/4

Different outcomes

Forward 1/4

Right 1/4

Figure 3.10. The different outcomes of the roll of two dice.

Scissors Congruence

As recreations, geometrical dissections have been around for thousands of years and the reader is undoubtedly familiar with the many forms of such puzzles.

The classical two-dimensional dissection puzzle typically involves a polygon which is to be partitioned by straight line segments into a finite number of polygonal subregions. The pieces are then rearranged using only rigid motions (isometries) to form some other specified polygon. If we ignore the boundaries of the pieces, we would say that the two polygons are *congruent by dissection, scissors,* or *jigsaw congruent.* More general shapes including curved boundaries and cuts allow for a wide variety of such puzzles.

To be set theoretically precise when partitioning point sets, we should consider every single point of the region to be partitioned, including those on the boundary and the cuts. A detailed discussion of this more precise form of decomposition and reassembly is given in the last section of this chapter on equidecomposability.

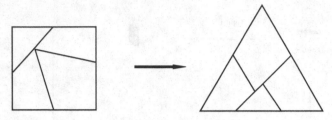

Figure 3.11. Square to equilateral triangle—four pieces.

The famous English puzzlist Henry E. Dudeney (1857–1930) published what may be the simplest dissection between two regular polygons. Though the pieces lack symmetry, the minimality (four pieces) makes the construction efficient (Figure 3.11).

Symmetry, if possible, adds elegance (Figure 3.12).

Figure 3.12. Octagon to square—five pieces.

American puzzlist Sam Loyd, mentioned in Chapter 2, published several dissection puzzles in the form of elaborate drawings (Figure 3.13).

Figure 3.13. Sam Loyd's sedan chair puzzle [Gardner 59, Figure 11].
(Reprinted with the permission of Dover Publications, Inc.)

Figure 3.14. Chair to square—two pieces.

The reader is asked to close up the chair and make a covered box by cutting the shape into the fewest possible pieces which can be reassembled to form a perfect square (Figure 3.14).

The ancient Chinese puzzle of Tangram can be found in most game stores. The origin of Tangram is not clear with one account, given by Sam Loyd, that it was invented by the God Tan over 4,000 years ago with each of the seven puzzle pieces respectively corresponding to the sun, the moon, Mars, Jupiter, Saturn, Mercury, and Venus. This account was later exposed as a historical spoof. The version found in stores today consists of a wooden (or plastic) square partitioned into seven pieces; one small square, two small isosceles right triangles, a medium-sized isosceles right triangle, two large isosceles right triangles and a parallelogram (Figure 3.15).

By rigid motion the pieces can be rearranged to form an incredible number of animals, shapes and other common objects (Figure 3.16).

The classical two-dimensional dissection puzzle requires a regular polygon be dissected into a finite number of pieces and reassembled to form another regular polygon. Is it possible that any polygon (regular or irregular) of given area is congruent by dissection to any other polygon having the same area? The following theorem answers in the affirmative.

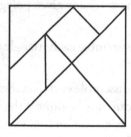

Figure 3.15. Tangram—seven plates of wisdom.

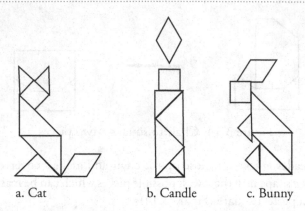

a. Cat   b. Candle   c. Bunny

Figure 3.16. Common Tangram shapes.

<u>Wallace-Bolyai-Gerwien Theorem</u>
· *Two polygons are congruent by dissection if and only if*
they have the same area. Specifically, any polygon is
congruent by dissection to a square of the same area.

Proof:

It is intuitive that two polygons which are congruent by dissection have the same area. The other direction of the proof requires that we show two polygons of the same area are congruent by dissection. To do so we need only to show that any polygon is congruent by dissection to a square of the same area. Then, to show two polygons of equal area are congruent by dissection, we need only consider that each can be dissected and reassembled to form the same square. Superimposing the two dissections on the single square gives us a dissection transforming the first polygon to the second.

Step 1. Partition the polygon into a finite set of triangles. (A proof that such a partition exists for any given polygon is given in [Eves 63, p. 238].)

Step 2. Dissect and reassemble each triangle to a rectangle as shown in Figure 3.17.

Step 3. Dissect and reassemble each rectangle to a square as shown in Figure 3.18. This can only be done if the length does not exceed four times the width. For a rectangle of excessive length, bisect and stack as necessary before converting to a square (Figure 3.19).

Step 4. Dissect and reassemble two squares to one (Figure 3.20).

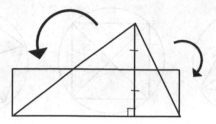

Figure 3.17. Triangle to rectangle.*

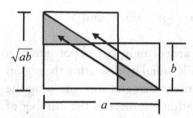

Figure 3.18. Rectangle to square.*

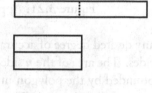

Figure 3.19. Bisecting and stacking as necessary.*

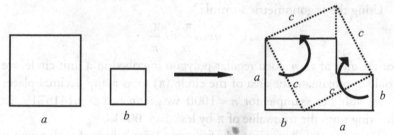

Figure 3.20. Two squares to one.*

Repeated application of Step 4 assembles all squares into a single square, as required.

The concept of dissection has mathematical applicability as well. For example, Step 4 above is a *proof* of the Pythagorean Theorem ($c^2 = a^2 + b^2$). Integral calculus uses dissections of shapes to find lengths, areas, and volumes by determining the limit of a sequence of successive approximations. For example, the area of a circle can be approximated, to

---

* Figures 3.17–3.20; [Wagon 85, p. 22].
(Reprinted with the permission of Cambridge University Press.)

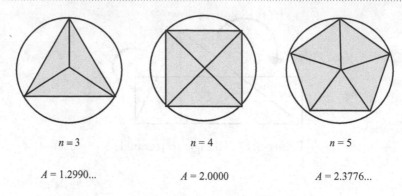

$n = 3$ $n = 4$ $n = 5$

$A = 1.2990...$ $A = 2.0000$ $A = 2.3776...$

Figure 3.21. Approximations for n = 3, 4, and 5.

any desired degree of accuracy, by the area of an inscribed polygon of n sides. The area of the n sided polygon is found by dissecting the region bounded by the polygon into n congruent triangles and summing the areas of the triangles. The approximation improves as the number of sides, n, increases. If the circle is a unit circle (radius one) then, using the fact that $A = \pi r^2 = \pi \cdot 1^2 = \pi$, we are able to approximate π to any desired degree of accuracy.

Successive approximations are shown in Figure 3.21 for n = 3, 4, and 5.

Using the trigonometric formula

$$A_n = n\cos\frac{\pi}{n}\sin\frac{\pi}{n}$$

for the area of an n sided regular polygon inscribed in a unit circle, we could approximate the area of the circle (π) to as many decimal places as we wish. For example, for $n = 1000$ we get an area of $3.1415719\ldots$ differing from the true value of π by less than .0001.

The table to follow gives A_n for some large values of n along with the error in using A_n as an approximation of π.

n	A_n	$\left|\pi - A_n\right|$
100	3.139525976	<.01
1,000	3.141571983	<.0001
10,000	3.141592447	<.000001
100,000	3.141592652	<.00000001
1,000,000	3.141592654	<.0000000001

Arc lengths, surface areas, and volumes can be approximated by similar methods. The approximations are found by dissecting and summing.

The desired sum, be it an arc length, surface area, or volume is defined as the limiting value of these ever improving approximations.

If we allow non-linear curves as boundaries in two-dimensional dissections, then we must set the rules of the game before proceeding. Which curves are admissible? The circle is a natural choice due to its symmetry, but there are no hard and fast rules.

Greg N. Frederickson's *Dissections: Plane and Fancy* is a classic collection of dissections, both two- and three-dimensional, some with curved boundaries. According to Frederickson [Frederickson 97, pp. 163–165], the following dissection of a circular table top (disc) by John Jackson in 1821 may be the earliest dissection of a curved figure. The disc is cut into eight pieces and rearranged to form two oval stools, with open handholds in the center of each. Much like the once popular TV show from the late 50s—*Name That Tune*—where contestants bid each other downward to determine who needed the fewest notes to identify a song, puzzlists tried to outdo each other by performing a specified dissection using fewer pieces. Loyd managed the *Disc to Ovals* dissection using only six pieces. In 1927, Dudeney published a four-piece dissection, giving credit to Loyd. All disc-to-oval dissections are done with circular cuts. (See Figure 3.22).

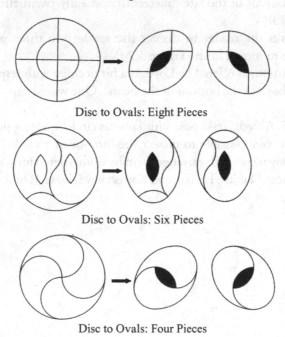

Disc to Ovals: Eight Pieces

Disc to Ovals: Six Pieces

Disc to Ovals: Four Pieces

Figure 3.22. Disc to ovals.

PROPOSITION—Show how to change a spade into a heart.

Figure 3.23. Sam Loyd's spade to heart puzzle [Gardner 59, Figure 59].
(Reprinted with the permission of Dover Publications, Inc.)

As was the case with polygonal dissections, both Loyd and Dudeney popularized curved boundary dissection puzzles in newspapers and other periodicals of the late nineteenth and early twentieth centuries (Figure 3.23).

Loyd asks the reader to dissect the spade into three pieces and reassemble to form a heart (Figure 3.24).

The solution, as offered by Loyd, is a bit peculiar with respect to the rounded shape of the bottom of the heart. One would expect more of a cusp.

In 1925, Alfred Tarski posed his famous circle squaring problem by asking if it were possible to dissect the interior of a circle (disc) into finitely many pieces and rearrange them by isometries to form a square of the same area. Dubins, Hirsch, and Karush have shown [Dubins, Hirsch,

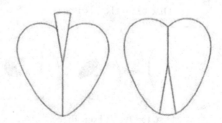

Figure 3.24. Spade to heart—Three pieces.

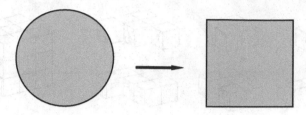

Figure 3.25. Disc to square—10^{50} pieces.

and Karush 63] that this is impossible if the dissection is restricted to scissors dissection. In 1990, Miklos Laczkovich of Budapest, Hungary proved [Laczkovich 90] that indeed it could be done by a more precise technique described in the following section on equidecomposability. In addition, he was able to show that it could be done via translations alone. The dissection proposed by Laczkovich uses the Axiom of Choice; therefore, the pieces are not explicitly defined. Laczkovich estimates that approximately 10^{50} pieces are required! (See Figure 3.25.)

The circle squaring problem of Alfred Tarski should not be confused with the classical problem of Greek mathematics of *squaring of the circle* or *quadrature of the circle*. Squaring the circle requires the straight edge and compass construction of a square equal in area to that of a given circle. In 1880 Ferdinand Lindemann (1852–1939) proved this to be impossible when he proved π to be a transcendental (non-algebraic) number. Given a circle of unit radius, it then becomes impossible to construct a square of the same area as the circle. The area of the square must equal π and its edge must be of length $\sqrt{\pi}$. Beginning with a line segment (radius) of length one, it is impossible to construct, by straight edge and compass alone, a line segment of transcendental length. The fact that π is transcendental implies that $\sqrt{\pi}$. is transcendental and the construction is impossible.

The simplest dissections of three-dimensional point sets are polyhedral dissections. A superb collection of such puzzles is Stewart T. Coffin's book *The Puzzling World of Polyhedral Dissections* [Coffin 98], now freely available at www.johnrausch.com/PuzzleWorld. Simple dissections of a $3 \times 3 \times 3$ cube are shown in Figures 3.26, 3.27, and 3.28.

Version 1 (Figure 3.26) is unusual in that all pieces are flat and the volume of the pieces increase in arithmetic sequence. Assembly is relatively easy.

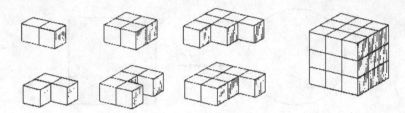

Figure 3.26. Cube dissection—Version 1.*

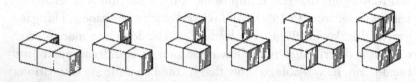

Figure 3.27. Cube dissection—Version 2.*

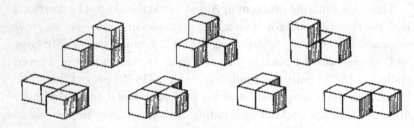

Figure 3.28. Cube dissection—Version 3.*

Version 2 (Figure 3.27), referred to as Mikusinski's Cube, after its originator J. G. Mikusinski, is slightly more difficult to assemble.

The seven piece dissection shown in Version 3 (Figure 3.28) is known as the Soma Cube and can be found in game and puzzle stores everywhere. Invented by Danish poet and inventor Piet Hein, it was marketed by Parker Brothers, Inc. in 1969 and popularized by Martin Gardner. Assembly is easy as there may be over one million possible ways to form the cube. Like the two-dimensional Tangram jigsaw recreation, many other shapes can be formed using all seven pieces (Figure 3.29 a, b, and c).

* (Reprinted with the permission of Stewart T. Coffin.)

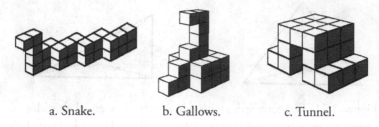

a. Snake.   b. Gallows.   c. Tunnel.

Figure 3.29. Common Soma shapes.

Some of the endless variety of three-dimensional jigsaw dissections appear in the photo in Figure 3.30, from the Coffin book.

It is natural to ask, and Hilbert did, if there is a three-dimensional analogy to the Wallace-Bolyai-Gerwien Theorem. Can any two polyhedra of equal volume be shown to be congruent by dissection? Hilbert suspected not. As the third problem on his famous list, he asked to show that there exist two tetrahedra of equal bases and altitudes which are not congruent by dissection.

Twenty-two-year-old German mathematician Max Dehn solved the problem before the list was presented by Hilbert in Paris; it was the first

Figure 3.30. Polyhedral dissection puzzles.
(Reprinted with the permission of Stewart T. Coffin.)

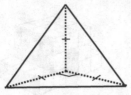 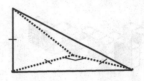

Figure 3.31. Two tetrahedra with equal base areas, equal altitudes, and equal volumes having different Dehn invariants [Yandell 02, p. 119]. (Reprinted with the permission of A K Peters, Ltd.)

of the list to be solved. As suspected by Hilbert, Dehn confirmed that in general, polyhedra of equal volumes are not congruent by dissection. His idea was to assign a number, the Dehn invariant, to a polyhedron based on the lengths of the edges and the angle between the faces. No amount of slicing and splicing could change this number for a given polyhedron. So, if two polyhedra (of equal volume) had different Dehn invariants, they would not be congruent by dissection. Dehn constructed two tetrahedra of equal volume having different Dehn invariants, thus confirming the impossibility of congruence by dissection (Figure 3.31).

If, in general, two polyhedra of equal volume are not congruent by dissection, how would it be possible to show, as the Banach-Tarski Theorem guarantees, that a solid of any given shape and volume can be decomposed and reassembled to form a solid of some other shape and/or volume? Doesn't the solution to Hilbert's third problem contradict the Banach-Tarski Theorem? Of course it must not. For one thing, Hilbert's third problem required that polyhedra be dissected by straight plane cuts into polyhedral pieces (scissors congruence), not allowing for other types of cuts and pieces. More importantly, the precise nature of the dissection of point sets has yet to be addressed, as we must account for each and every point of the set, including edges and boundaries. Strangely, it is this very precision which allows the Banach-Tarski Theorem to work its magic.

Equidecomposability

We have been a bit cavalier in the dissection of point sets as discussed in the previous section. Bear in mind that point sets are just that—sets of points; so, we must pay careful attention to each and every point of the set being dissected, including the edges and boundaries. Recreational

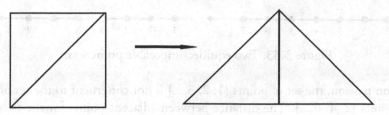

Figure 3.32. Square to isosceles right triangle—two pieces.

dissections typically ignore such matters but mathematical precision requires careful attention.

Consider the problem of dissecting a unit square (interior and boundary) into a finite number of pieces and forming an isosceles right triangle. As a jigsaw dissection, it is trivial and can be done in only two pieces (Figure 3.32).

But to be mathematically precise we must answer the following:

1. To which piece do we assign the points along the diagonal cut of the square? The diagonal can only be used for one of the hypotenuses of the two triangular pieces.

2. When the two triangular pieces are spliced to form the large right triangle, what is happening along the seam? Apparently two edges of the square on the left are spliced. Are points not lost in the process?

If we cannot satisfactorily answer these questions, then we are not able to transform the square into the isosceles right triangle in such a way that there is a one-to-one correspondence between points of the square and points of the right triangle. As we shall see, it is possible to transform the square into the isosceles right triangle by finite dissection, taking all points into account. We say the square and the isosceles right triangle are *equidecomposable* or *piecewise congruent*, meaning that one can be decomposed into a finite number of pieces and reassembled to form the other, *taking into account all points including edges and boundaries.*

Think of equidecomposability as a more precise form of congruence by dissection.

To be clear on equidecomposability, we begin by considering point sets on the real number line. The sets of points {1, 2, 3,...} and {3, 4, 5,...} are congruent because the first can be shifted to the right two units (an isometry) and be made to coincide with the second. For

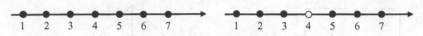

Figure 3.33. Two equidecomposable point sets.

comparison, the set of points {1, 2, 3,...} is not congruent to the set of points {2, 4, 6,...}. The distance between adjacent points for the two sets differs by a factor of two and the first can not be shifted to coincide with the second. These two sets have the same cardinality of $\aleph_0$, but they are not congruent.

Now consider the following two point sets: {1, 2, 3,...} and {1, 2, 3} ∪ {5, 6, 7, . . .}. So, the first point set can be thought of as points on the number line at every counting number and the second point set is the same, except for a *hole* at 4. The two sets are not congruent (Figure 3.33).

However, the two point sets are equidecomposable. If the set on the right is decomposed into the two sets {1, 2, 3} and {5, 6, 7,...} then the set {5, 6, 7,...} can be shifted to the left one unit (an isometry) to plug the hole at 4 and form the set {1, 2, 3,...}. We could refer to this as *shifting from infinity*. Equivalently we could start with the set on the left {1, 2, 3,...} and partition it into the two sets {1, 2, 3} and {4, 5, 6,...} Then the set {4, 5, 6,...} could be shifted to the right one unit forming the set {5, 6, 7,...}. We call this *shifting to infinity* and say that the 4 has been *absorbed* in doing so. This creates a hole at 4.

To clarify absorption by shifting to infinity, we turn to David Hilbert's often told story of *Hilbert's Hotel*—the hotel with infinitely many rooms. The hotel is full and a traveler arrives at the front desk, inquiring about a vacancy. The proprietor solves the problem by shifting all guests (to infinity) up one room. So the guest occupying room 1 is moved into room 2, the guest in room 2 is moved to room 3, and so on. The process vacates room 1, allowing for the traveler at the front desk to be absorbed as a guest.

Even if an infinite number of new customers arrive at the full hotel, accommodation is possible. The proprietor shifts the guest in room 1 to room 2, the guest in room 2 to room 4, the guest in room 3 to room 6, the guest in room 4 to room 8, and so on. Now all odd numbered rooms are free and the infinite number of new guests can be accommodated.

Lawrence M. Lesser of the University of Texas at El Paso cleverly adapts the concept as a parody of the Eagles' song "Hotel California" (Don Felder, Don Henley, Glenn Frey). He calls it "Hotel Infinity."

Hotel Infinity

On a dark desert highway – not much scenery
Except this long hotel stretchin' far as I could see.
Neon sign in front read "No Vacancy,"
But it was late and I was tired, so I went inside to plea.

The clerk said, "No problem. Here's what can be done—
We'll move those in a room to the next higher one.
That will free up the first room and that's where you can stay."
I tried understanding that as I heard him say:

CHORUS: "Welcome to the HOTEL INFINITY—
Where every room is full (every room is full)
Yet there's room for more.
Yeah, plenty of room at the HOTEL INFINITY—
Move 'em down the floor (move 'em down the floor)
To make room for more."

I'd just gotten settled, I'd finally unpacked
When I saw 8 more cars pull into the back.
I had to move to room 9; others moved up 8 rooms as well.
Never more will I confuse a Hilton with a Hilbert Hotel!

My mind got more twisted when I saw a bus without end
With an infinite number of riders coming up to check in.
"Relax," said the nightman. "Here's what we'll do:
Move to the double of your room number: that frees the
odd-numbered rooms."

(Repeat Chorus)

Last thing I remember at the end of my stay—
It was time to pay the bill but I had no means to pay.
The man in 19 smiled, "Your bill is on me.
20 pays mine, and so on, so you get yours for free!"

The above ideas of shifting to infinity and absorption are now used to show that a circle is equidecomposable to a circle of the same radius with one point removed. It can be done in two pieces (Figure 3.34)!

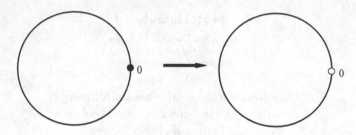

Figure 3.34. Circle to circle missing one point—two pieces.

Let the circles each be of radius 1. Beginning with the circle on the left, designate the point to be removed as 0. Travel a distance of one unit counterclockwise from 0 and mark it as 1. Travel an additional unit counterclockwise and call it 2. Continue in this manner. The first few points so designated are shown in Figure 3.35.

Considering that the circumference of the circle is irrational (2π), we could continue marking points around the circle in this manner ad infinitum with no two points ever coinciding.

Call the set of all such points A. That is, $A = \{0, 1, 2, 3, \ldots\}$. Let B be the set of the points on the circumference and not in A. So, we can think of the complete circle as being the union of A and B. Now rotate all points in A one step counterclockwise. So, 0 is moved up to 1, 1 is moved up to 2, 2 is moved up to 3, and so on. We are shifting A to infinity one step and leaving B unchanged. In doing so, we are vacating the point at 0 because the point at 0 is being absorbed into the

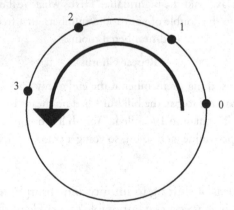

Figure 3.35. Shifting to infinity.

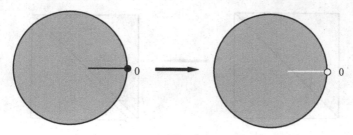

Figure 3.36. Disc to disc missing line segment—two pieces.

circle. After shifting, we are left with the circle with one point removed. Equivalently, we could start with the circle on the right and shift from infinity, plugging the hole.

In similar fashion we show that a disc is equidecomposable to a disc missing a line segment (Figure 3.36).

It is best for our purposes to assume the segment has only one endpoint. Shifting to infinity allows the line segment shown in the left disc to be absorbed. The argument is identical to that involving the circle with the missing point. Or, starting with the disc on the right, we could shift from infinity to fill the gap (Figure 3.37).

Now back to our original problem of showing that a square is equidecomposable to an isosceles right triangle of the same area.

The square is cut diagonally with all points on the diagonal belonging to the upper left piece, as indicated by the solid and dotted lines (Figure 3.38).

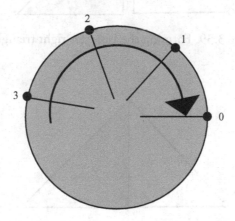

Figure 3.37. Shifting from infinity.

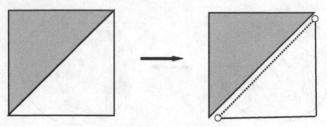

Figure 3.38. First cut.

The two small right triangles are then brought together as shown to form the large isosceles right triangle (Figure 3.39).

Unfortunately, we now face two obvious problems. Only one of the two vertical line segments in Figure 3.39 can serve as the altitude of the isosceles right triangle. Also, we are missing the upper left side of the triangle. A partial remedy would be to move one of the two candidates for the altitude, say the one on the right, to the position of the missing side, as shown. Then the two small right triangles could be spliced together without placing points on top of points. We end up with the Figure 3.40.

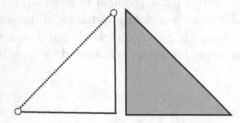

Figure 3.39. Forming the isosceles right triangle.

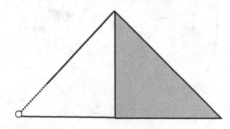

Figure 3.40. Isosceles right triangle—Almost!

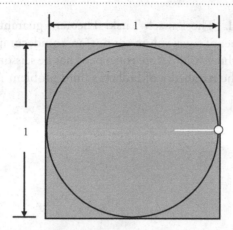

Figure 3.41. Filling the gap by shifting from infinity.

We are missing a set of points forming a line segment with one endpoint. Since the original square had a side length of 1, the length of the missing segment is $\sqrt{2} - 1$ or approximately .4.

It seems so insignificant! Too bad we can't just remove it from the original square and insert it where needed on the side of the isosceles right triangle. Surely it wouldn't be missed, right? Actually it wouldn't be missed at all and this is exactly what we are going to do! If we remove it from the interior of the original square as shown in Figure 3.41, then we will have created a gap. But the gap is only temporary as it can be filled by the shifting from infinity process discussed earlier. We could inscribe a circle inside the original square and remove our line segment of length $\sqrt{2} - 1$ as shown. We could then fill in the gap as we have done previously.

From this point on, we will be precise in our dissections and reassemblies, accounting for every last point of the original point set and being careful that points do not overlap when the pieces are reassembled. That is, we desire equidecomposability, not just scissors congruence. As we have seen, these two types of piecewise congruencies are not equivalent. Equidecomposability does not imply scissors congruence as we have already seen that a circular disc is equidecomposable to a square of the same area, but the two sets are not scissors congruent. However, the reverse implication does hold. Scissors congruence implies equidecomposability as the boundaries (edges) of the pieces can be absorbed during reassembly. And, this applies to three-dimensional point sets as well. The Banach-Tarski Theorem guarantees any two polyhedra, whether of equal volume or not, are equidecomposable. Yet, even polyhedra whose volumes are equal need not be scissors congruent, as is the case of the tetrahedra of Hilbert's third problem.

4 Baby BTs

Paradox likes contradiction with exits.
—Mason Cooley

Paradox breeds paradox!

Georg Cantor defined an infinite set as one which could be put in one-to-one correspondence with a proper subset of itself. This would appear to some as contradicting Euclid's common notion of the whole being greater than the part. If the very definition of infinite set is paradoxical, it should be no surprise that additional paradoxes involving infinite sets will follow. Notable examples, discussed in the preceding two chapters, include the Burali-Forti Paradox, Cantor's Paradox, and Russell's Paradox. More are presented in this chapter.

Paradoxes of this sort are Type 1 paradoxes, appearing contradictory, even absurd, yet are true in actual fact. Let's not confuse these with Type 2 paradoxes which are sleights and are the result of a fallacious argument or misdirection. Chapter 2 included many Type 2 jigsaw paradoxes involving the rearranging of shapes in such a way that something was apparently gained or lost. In all cases, those paradoxes could be resolved by exposing the sleight. In comparison, Type 1 paradoxes are resolved once their truth is understood.

This chapter will present several Type 1 paradoxes, each similar in some respects to the Banach-Tarski Paradox. Most are geometrical and all involve a rearrangement where something is either gained or lost. Each can be thought of as a precursor to the most stunning of such paradoxes—the Banach-Tarski Theorem.

It might suffice to define an infinite set as a set which is not finite. The problem with this negative definition is that it tells us what an infinite set is not, as opposed to what it actually is. It is relatively non-descriptive and adds no information of value. Instead we choose the definition adopted by Cantor.

Definition of Infinite Set

A set is infinite if there exists a one-to-one correspondence between its elements and the elements of one of its proper subsets.

Let's be clear about *one-to-one correspondence*. A mapping of one set to another is called one-to-one if distinct elements of the first set are mapped to distinct elements of the target set. For example, the mapping $x \mapsto x^3$ is one-to-one, whereas the mapping $x \mapsto x^2$ is not one-to-one. (Consider $1 \mapsto 1$ and $-1 \mapsto 1$.) A one-to-one mapping is called an *injection*. A mapping (whether or not it is one-to-one) is *onto* a second set if every element of the second set is mapped to by at least one element of the first set.

For example, the mapping $x \mapsto x^2$ is a mapping of the real numbers onto the non-negative real numbers. An onto mapping is referred to as a *surjection*. A mapping that is both one-to-one and onto is called a *bijection* and is equivalent to there being a one-to-one correspondence between elements of the sets.

The definition of an infinite set adopted by Cantor is blatantly paradoxical, almost taunting the reader to make some sense out of it. It is not self contradictory. It suggests a nesting process of infinite regression as shown in Figure 4.1.

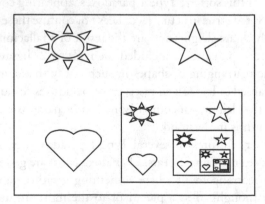

Figure 4.1. Infinite regression.

Formally adopted by Cantor, this definition of infinite set was considered by several of Cantor's predecessors. Galileo Galilei (1564–1642), the Italian physicist and mathematician, perhaps best known for inventing the telescope and his defense of the Copernican or heliocentric center of the universe, may have been the first to consider the peculiar nature of infinite collections in terms of proper subsets. While under house arrest by the Inquisition and just four years before his death, Galileo wrote *Two New Sciences* (1638) a philosophical and mathematical treatise in the form of a dialogue between the intelligent Salviate, the layman Sagredo, and the simpleton Simplicius. Though rich with academic content, its purpose may have been to criticize the Inquisitors as represented by Simplicius. Salviate defines a one-to-one correspondence between the set of all integers and the set of perfect squares of integers and concludes [Galileo Galilei 74, pp. 40–41]

> . . . it must be said that square numbers are as numerous as all numbers, . . .

> . . . that the multitude of squares is not less than that of all numbers, or is the latter greater than the former.

In comparison to Galileo's discussion of infinite sets of integers, Bernhard Bolzano (1781–1848) considered dense points sets, such as intervals on the real number line. Bolzano, a Czechoslovakian priest whose mathematical work on infinity was overlooked by his contemporaries, completed *Paradoxien des Unendlichen* (*Paradoxes of the Infinite*) just eighteen days before his death. It was published posthumously in 1850. By considering the function $y = 2x$, he established a one-to-one correspondence between the points of the line segment [0, 1] and points of the line segment [0, 2]. Similar to Galileo's discussion of infinite sets of integers, Bolzano showed these two intervals were *equal* in the sense of cardinality, despite their inequality with respect to length. It is as if the interval [0, 1], a subset of [0, 2], can be stretched, by simply rearranging the position of each of its points, to create an interval of twice its original length.

In *Paradoxien des Unendlichen*, Bolzano discusses this *equality-inequality* paradox:

> . . . many of them (intervals) are *greater* (or *smaller*) than some other in the sense that the one includes the other as a part of itself (or stands to the other in the relation of part to whole). Many consider this as yet another *paradox*, and indeed, in the

eyes of all who define the infinite as that which is incapable of increase, the idea of one infinite being greater than another must seem not merely paradoxical, but even downright *contradictory*.

With regard to the existence of a one-to-one correspondence, Bolzano writes

. . . in spite of their entering symmetrically into the above relation (one-to-one correspondence) with one another, the two sets can still stand in a relation of inequality, in the sense that the one is found to be a whole and the other a part of the whole.

More will be said of this in the stretching section of this chapter.

Richard Dedekind (1831–1916), the *last student of Gauss*, first met Georg Cantor in 1874 while the two were on vacation in Interlaken, Switzerland. Dedekind was critical of Kronecker's objection to the infinite, thus aligning himself with the views of Cantor. The two would become friends, with Dedekind remaining sympathetic to Cantor's set theory. Though known primarily for his definition of irrational numbers in terms of *Dedekind cuts*, he made significant contributions to number theory, and set theory, using the definition for infinite set as adopted by Cantor. Today this definition is called the *Dedekind definition* of infinite set.

The following definition of infinite set is equivalent to the Dedekind definition and appears less paradoxical.

Alternate Definition of Infinite Set

A set is infinite if there exists a one-to-one mapping (injection) from the set of counting numbers into the given set.

Using the Axiom of Choice it is easy to show the equivalence of the two definitions by showing each implies the other. For this purpose, we refer to the first definition as the *Dedekind definition* and the second (alternate) definition as the *injection definition*.

To show the Dedekind definition implies the injection definition, we begin by assuming a given set S is infinite in the sense that there is a bijection between S and a proper subset S_1 (Figure 4.2). It would follow that S_1 is also infinite (having the same cardinality as S) and therefore can be put into one-to-one correspondence with one of its

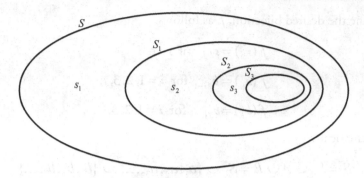

Figure 4.2. Dedekind definition implies injection definition.

proper subsets, S_2. Proceeding in this manner, we form a countably infinite sequence of sets each being a proper subset of its predecessor; $S \supset S_1 \supset S_2 \supset S_3 \supset \dots$.

Using the Axiom of Choice, choose

$$s_1 \in S - S_1, s_2 \in S_1 - S_2, \dots s_i \in S_{i-1} - S_i, \dots$$

The desired injection is the association of s_i with each counting number i.

To show the injection definition implies the Dedekind definition, we begin by assuming S is infinite in the sense that there exists a one-to-one mapping, or injection, from the set of counting numbers into S. If S is countable it is easy to show the Dedekind definition applies. (Let $S = \{s_1, s_2, s_3, \dots\}$. A simple one-to-one correspondence between S and one of its subsets would be given by $f(s_i) = s_{2i}$, a bijection from S to the even indexed elements of S.) If S is uncountable we have a bit more work to do. Let the injection be defined by the function $f(i) = a_i$ where i is a counting number and $a_i \in S$, $i = 1, 2, 3, \dots$. Being one-to-one, we understand $a_i \neq a_j$, for $i \neq j$. Let $A = \{a_1, a_2, a_3, \dots\}$. We will show $n(S) = n(S - A)$, establishing that S has the same cardinality as one of its proper subsets. Having the same cardinality, there would exist a bijection between the two.

The set $S - A$, being infinite (and uncountable), must contain a countable subset, say $B = \{b_1, b_2, b_3, \dots\}$. Let $S^* = S - (A \cup B)$. Note the following:

$$S = S^* \cup A \cup B = S^* \cup \{a_1, a_2, a_3, \dots\} \cup \{b_1, b_2, b_3, \dots\}$$
$$S - A = S^* \cup \{b_1, b_2, b_3, \dots\}$$

Define the desired bijection, f, as follows:

$$f(s^*) = s^* \quad \text{if } s^* \in S^*$$

$$f(a_i) = b_{2i-1} \quad \text{for } i = 1, 2, 3, \ldots$$

$$f(b_i) = b_{2i} \quad \text{for } i = 1, 2, 3, \ldots$$

Schematically,

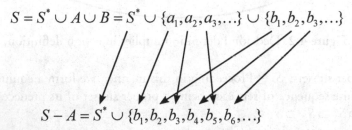

$$S = S^* \cup A \cup B = S^* \cup \{a_1, a_2, a_3, \ldots\} \cup \{b_1, b_2, b_3, \ldots\}$$

$$S - A = S^* \cup \{b_1, b_2, b_3, b_4, b_5, b_6, \ldots\}$$

This establishes the one-to-one and onto correspondence as required.

The Baby BTs are presented in the following six sections of this chapter:

1. Shifting to Infinity
2. Stretching
3. Cantor Dust
4. The Vitali Construction Paradoxes
5. Stewart's Hyperwebster Dictionary
6. The Sierpiński-Mazurkiewicz Paradox

Shifting to Infinity

The fact that an infinite point set can be put into one-to-one correspondence with one of its proper subsets allows for apparent and paradoxical gains and losses by shifting to and from infinity as presented in the previous chapter. Shifting to infinity absorbs points and shifting from infinity creates points. Are we not violating some law of conservation when we show the *part is as big as the whole*? This

Figure 4.3. More squares than dots—by inspection.

paradox, allowed by the Dedekind definition of infinite set, may be the seminal paradox of those which follow and we best resolve it now before moving on.

We learn from experience and we have hands on experience with finite sets. We have two eyes, ten fingers, and live on a planet with approximately 6.4×10^9 human inhabitants. The mass of our planet is approximately 6×10^{24} kilograms and we are 4×10^{13} kilometers from the nearest star system—Alpha Centauri. Though large, these numbers are finite and we are clear what it means for one finite number to be greater than (or less than) another. We have, however, no tangible experience with infinite collections and we must be very precise when we use such terms as "greater than" and "less than" with respect to infinite sets.

Consider the simple example of finite collections of squares and dots as shown in Figure 4.3.

We quickly determine there are more squares than dots by noting the shift on the left and the fact that both rows stop at the same point on the right. We naturally pair up the squares and dots (vertically) and note there are two more squares than dots. There is no need to count. Experience tells us our conclusion is valid. We know there can not be a one-to-one correspondence between the squares and the dots as this would violate the *pigeon hole principle*.

Now consider infinitely many squares and dots as shown in Figure 4.4.

As before, we tend to pair up the squares and dots vertically. Intellectually we know both sets are infinite; yet, we may believe the two collections stop together *at infinity*. When you think about it, the phrase, *at infinity*, is meaningless. The squares appear more numerous

Figure 4.4. More squares than dots?

Figure 4.5. Shifting to infinity preserves cardinality.

than the dots because of the two extra squares on the left. Clearly the totality has not changed as we can still pair the squares and dots in one-to-one fashion as illustrated in Figure 4.5.

So, what actually happens when the countably infinite point set $\{1, 2, 3, \ldots\}$ is shifted (to infinity) two steps to the right yielding $\{3, 4, 5, \ldots\}$? Are points not lost? In fact the points named 1 and 2 are lost but *the totality of points does not change*. Mathematically we write $\aleph_0 - 2 = \aleph_0$ or more generally $\aleph_0 - n = \aleph_0$ for all counting numbers n. Similarly, if the point set $\{3, 4, 5, \ldots\}$ were shifted (from infinity) to the left two steps yielding $\{1, 2, 3, \ldots\}$ we would be gaining the two points 1 and 2 without changing the totality, or cardinality of the original point set. In general, we write $\aleph_0 + n = \aleph_0$. For some this may be a tough pill to swallow as it appears we now have some new form of arithmetic. Precisely! The rules of transfinite arithmetic are different from those of finite arithmetic (Figure 4.6).

In Chapter 3, we showed that a circle missing a single point is equidecomposable to a complete circle of the same radius. The shifting to and from infinity technique used there is precisely the same as the one we are discussing now. Instead of thinking of the point set $\{1, 2, 3, \ldots\}$ as an infinite set of points on a straight line, we could consider the points spaced one unit apart on the circumference of a circle of radius one. Beyond this, the concept is the same.

Similar *shifting to infinity* paradoxes exist in two- and three-dimensional space as well. We can shift half of the Cartesian plane two units to the right and obtain a two-dimensional point set which is a proper subset of the original half plane (Figure 4.7). We have lost some points. In fact

$\aleph_0 + n = \aleph_0$ (See explanation above.)

$\aleph_0 + \aleph_0 = \aleph_0$ (Consider $n(\{1,3,5,\ldots\} \cup \{2,4,6,\ldots\}) = n(\{1,2,3,\})$.)

$n \times \aleph_0 = \aleph_0$ (Proceed inductively from above.)

Figure 4.6. Some strange laws of transfinite arithmetic.

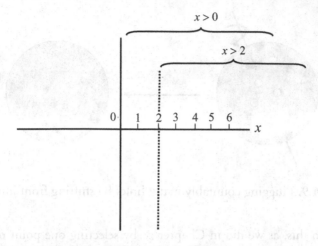

Figure 4.7. Shifting the half plane to infinity.

we have lost an uncountable infinity of such points. Yet, the totality (or cardinality) of points in the shifted set must equal the number of points in the original half plane. The one-to-one correspondence of points should be clear.

The paradox subsides when we accept the fact that totality is not changing, despite the fact points are being added or removed. Such is the nature of infinite point sets.

We give one more example of shifting to infinity as it will be a significant component of our proof of the Banach-Tarski Theorem. It is a simple matter to show that the surface of a sphere with one point removed is equidecomposable to the complete sphere. Simply consider the missing point as belonging to a circle on the sphere's surface and shift from infinity, as in Chapter 3, to plug the hole (Figure 4.8).

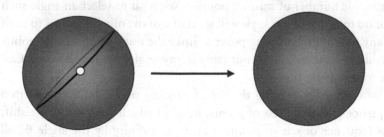

Figure 4.8. Plugging the hole by shifting from infinity.

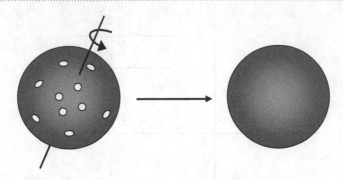

Figure 4.9. Plugging countably many holes by shifting from infinity.

We do this, as we did in Chapter 3, by selecting one point on the circle's circumference and creating additional points on the circumference by advancing a distance which is some irrational multiple of the circle's circumference. This can be done in uncountably many ways as there are uncountably many irrational numbers. (In Chapter 3, we advanced around the circle a distance of one unit, which is an irrational multiple of the circle's circumference of 2π.) We then shift back from infinity, as in Chapter 3, to plug the hole on the sphere's surface.

By the exact same technique, we can show that the surface of a sphere missing a countable number of points (holes) is equidecomposable to the complete sphere.

Begin by choosing an axis of rotation which does not intersect any of the holes. With uncountably many points on the surface of the sphere and only countably many holes, there are uncountably many ways to choose the axis (Figure 4.9).

Each missing point (hole) traces a circle as the sphere is rotated. So there are at most a countable number of circles being traced by the countable number of missing points. We wish to select an angle such that no multiple of the angle will send one of the missing points to itself or any other of the missing points. Since the number of missing points is countable, there are uncountably many angles to choose from. Call the special angle θ.

By repeatedly rotating the set of missing points by θ, we form a sequence of disjoint sets of points, none of which coincide. If we shift this sequence of sets of points back from infinity by the angle θ, all holes are simultaneously plugged.

Being able to plug these holes is critical in our proof of the Banach-Tarski Theorem. This procedure will come near the end of the proof and will allow us to create two spheres (surfaces) from one. It will then be a simple matter to extend the paradox from the spherical surfaces to solid balls.

Stretching

Ask someone, "Are there more points on a line segment of length two than there are on a line segment of length one?" Answers will vary. The correct answer is that the number of points on each segment is equal to the power of the continuum, c; therefore each segment has the same transfinite number of points. Somehow, we are able to independently rearrange the points of the line segment of length one and *stretch* the segment to double its length. As noted by Bolzano, the equality of cardinality (content) exists despite the inequality of length.

As before we can attribute the paradox to that which we have direct experience. It is natural to perceive a point as a period on this page or a grain of sand, each, to be sure, quite small, yet of finite measure. The period on this page may be a small disc of diameter .1 mm and a grain of sand would be a three-dimensional solid of small finite volume. We understand that combining finite numbers of these periods or grains of sand yield shapes of finite size. Combining infinite quantities yield shapes that are infinitely large. But in mathematical actuality, points have zero measure and are to be thought of as locations. Being infinitesimally small, the rules of finite arithmetic do not apply. It is as hard to let go of our old habits in working with infinitesimally small quantities as it is to accept the laws of transfinite arithmetic which apply to infinitely large cardinalities.

In the pages to follow, we will consider stretching paradoxes of one, two, and three dimensions. Can a point set be stretched to a point set of higher dimension? The answer may surprise you, as it did Cantor. The Baby BTs in this section involve stretches (or compressions) where points are paradoxically gained (or lost).

Examples of one-dimensional stretches are given in Figure 4.10. Each is a *graphical proof* of the fact that one-dimensional points sets can consist of the exact same number of points, despite their inequality with respect to their length. This is the equality-inequality paradox as noted by Bolzano.

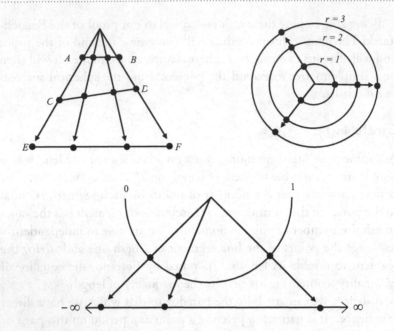

Figure 4.10. One-dimensional point sets having the cardinality of the continuum.

The figure in the upper left shows the one-to-one correspondence of the point sets associated with line segments *AB*, *CD*, and *EF*. Each set contains the same number of points, despite their varying lengths. So, in a sense, the points of line segment *AB* can be rearranged to form line segments *CD* and *EF*. We are somehow stretching *AB* to conform to *CD* and *EF* without increasing the cardinality of the point set *AB*.

The three concentric circles shown in the upper right illustrate the same phenomenon. Circles of any given radius have the same number of points on their circumference.

The illustration on the bottom was given in Chapter 1. The semicircular arc, of length one, is of finite length and the real number line is of infinite length. Yet, there are clearly the same number of points on each. All point sets illustrated contain the same number of points. The number is *c*, the power of the continuum.

A more precise way to guarantee that two sets are of equal cardinality is to exhibit the bijection (one-to-one and onto) which maps each point of one set to a unique point of the other. This is easily done for the examples illustrated. If *x* is a member of the set of all points between $x = 0$ and $x = 1$ (we write $x \in [0, 1]$) then $2x \in [0, 2]$. All line segments of finite length can be shown in one-to-one correspondence in this manner. In similar fashion, if (x, y) is a point on a circle of radius one centered at $(0, 0)$ then $(2x, 2y)$ is a point on a circle of radius two centered at $(0, 0)$. The argument extends to circles of any given radius.

A variety of bijections exist establishing the one-to-one correspondence between $(0, 1)$ and $(-\infty, \infty)$ Bijections from $(0, 1)$ to $(-\infty, \infty)$ include

$$x \mapsto \frac{2x - 1}{2x^2 - 2x} \quad \text{and} \quad x \mapsto \tan\left(\pi x - \frac{\pi}{2}\right).$$

Bijections from $(-\infty, \infty)$ to $(0, 1)$ include

$$x \mapsto \frac{|x| + x + 1}{2|x| + 2} \quad \text{and} \quad x \mapsto \frac{1}{2} + \frac{1}{\pi}\arctan x.$$

The stretching paradox can only be resolved once we accept the fact that a point has zero measure, contrary to our real world experience with such points as text periods and grains of sand. It may help to think of a point as a location, as opposed to an object and then ask yourself one more time, "Are there more locations on a line segment of length one than on a line segment of length two?"

Consider the two linear point sets $\{x : 0 < x < 1\}$ and $\{x : 0 \le x \le 1\}$. Can the first be stretched (magnified) in such a way as to coincide with the second? We denote the first set as the open interval $(0, 1)$—endpoints excluded—and the second set as the closed interval $[0, 1]$—endpoints included (Figure 4.11). They are equal in length and differ by only the two endpoints. We would expect the cardinality of both sets to be the same as that of the continuum—c.

The stretch, if possible, would be minimal in comparison to the others. But how can we show it? It is harder than it looks to exhibit a bijection from one set to the other. So how can we be certain that the cardinalities are the same and the elements of each set can be put in one-to-one correspondence?

0	1	0	1

Figure 4.11. A slight stretch?

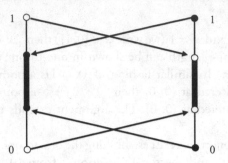

Figure 4.12. Applying the Schröder-Bernstein Theorem.

We can use the Schröder-Bernstein Theorem, mentioned in Chapter 3. A version of this theorem is used to prove the Banach-Tarski Theorem and we prove this version in Chapter 5. The theorem states that if there exists an injection (one-to-one) function from each of two sets into the other, then there exists a bijection from one set to the other and the sets are of the same cardinality. As Figure 4.12 shows, it is easy to establish an injection from (0, 1) into [0, 1] and in the reverse direction by simple stretching (or rather compressing) as previously discussed.

We see the two injections and the two point sets are necessarily of the same cardinality. So we should be able to find a bijection from one to the other. The trick involves keeping some points fixed and shifting others. We present a bijection from [0, 1] to (0, 1).

Consider $\left\{0, 1, \dfrac{1}{2}, \dfrac{1}{3}, \dfrac{1}{4}, \ldots\right\} \subseteq [0, 1]$ and let $A = [0, 1] - \left\{0, 1, \dfrac{1}{2}, \dfrac{1}{3}, \dfrac{1}{4}, \ldots\right\}$.

So $[0, 1] = A \cup \left\{0, 1, \dfrac{1}{2}, \dfrac{1}{3}, \dfrac{1}{4}, \ldots\right\}$.

Define the bijection from [0, 1] to (0, 1) as shown in Figure 4.13.

$$[0, 1]: \quad A \quad \cup \quad \{\ 0,\ 1,\ \dfrac{1}{2},\ \dfrac{1}{3},\ \dfrac{1}{4},\ \ldots\}$$

$$\downarrow \qquad\qquad \downarrow\quad \downarrow\quad \downarrow\quad \downarrow\quad \downarrow$$

$$(0, 1): \quad A \quad \cup \quad \{\ \dfrac{1}{2},\ \dfrac{1}{3},\ \dfrac{1}{4},\ \dfrac{1}{5},\ \dfrac{1}{6},\ \ldots\}$$

Figure 4.13. A bijection from [0, 1] to (0, 1).

Other bijections exist similar to the example in Figure 4.13 that keep some points fixed and shift others. Strangely, none is continuous. There is no bijection which will stretch or magnify $(0, 1)$, be it ever so slightly, to transform it to $[0, 1]$ in a continuous way. It takes a significant reshuffling of the points to do the job. It's strange how $(0, 1)$ can be stretched continuously by a bijection to twice its size, or even the entire real number line, without a problem. Yet to find a bijection which extends it by only two points requires a clever discontinuous mapping. This can not be done continuously by magnification; therefore the stretch, in the sense of stretching a rubber band, is impossible. The reason for the difficulty stems from a theorem (proof omitted) which states that compact (closed and bounded) sets are mapped by continuous functions into compact sets. So, it would be impossible to take the compact point set $[0, 1]$ and map it continuously (compress it) to the open set $(0, 1)$. The reverse (stretching) direction is equivalently impossible.

In similar fashion bijections exist from $[0, 1]$ to $[0, 1)$ and $[0, 1]$ to $(0, 1]$. The bijections involve a significant shuffling of points and are not continuous, despite the minimality of the extension.

In summary, all interval point sets, be they closed, open, of finite or infinite length possess the same number of points and this number is denoted by c, the power of the continuum.

Two-dimensional stretching paradoxes are similar to the one-dimensional stretches. Several are illustrated in Figure 4.14 with a

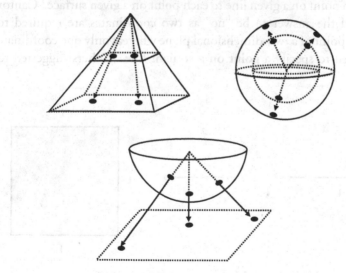

Figure 4.14. Two-dimensional stretches.

graphical proof that points of these sets can be put into one-to-one correspondence.

The figure in the upper left of Figure 4.14 suggests that all point sets in the form of a square (including the interior) are of the same cardinality. The upper right illustration of concentric spherical surfaces is meant to show that all such surfaces consist of the same number of points, despite the fact they are of different measures. The bottom illustration shows that the points of a hemispherical surface can be put into one-to-one correspondence with the infinite two-dimensional plane.

How does the cardinality of these two-dimensional point sets compare to c, the cardinality of the one-dimensional point sets discussed earlier? One might expect a two-dimensional space to contain more points than a one-dimensional space as it is, in a sense, more extensive. Is it conceivable that the points of a one-dimensional point set, such as a line segment, can be put into one-to-one correspondence with a two-dimensional point set? A line has no *width* and it seems unnatural that a given area and a given line could contain the same number of points.

Cantor asked himself this very question. Are there more points on the plane than on a line? In general, how does the dimension of a point set affect the cardinality of that point set? The simplest case to investigate would be that of a line segment of length one and a square (including interior) of side one (Figure 4.15).

In 1874, Cantor wrote to Dedekind asking if it would be possible to map each point on a given line to each point on a given surface. Cantor suspected the answer to be "no" as two coordinates are required to specify a point of a two-dimensional plane whereas only one coordinate is required to specify a point on a straight line. Others suggested to

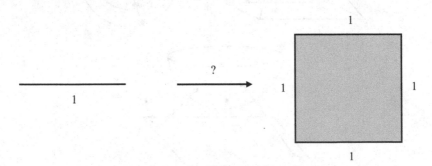

Figure 4.15. Line segment to square.

Cantor that the matter wasn't worth pursuing as it would be impossible to find a one-to-one correspondence between the points of sets of different dimension.

Cantor was astonished to discover that there was indeed such a correspondence. In 1897, he wrote back to Dedekind, in French [Aczel 00, p. 119], "Je le vois, mais je ne le crois pas!" (I see it but I don't believe it!) Once again Cantor felt compelled to accept results that were difficult to comprehend. Dedekind wrote back to Cantor, cautioning him that his results would not be well received. Publishers such as Crelle's Journal were hesitant to accept articles submitted by Cantor as there were concerns of possible errors.

There are numerous ways of showing the one-to-one correspondence, including the Schröder-Bernstein technique. Here we choose to give a simple bijection between the point set $(0, 1] = \{x : 0 < x \leq 1\}$ and the two-dimensional point set $\{(x, y) : 0 < x \leq 1, \ 0 < y \leq 1\}$.

As discussed in Chapter 1, every real number between 0 and 1 can be represented by a non-terminating decimal fraction; i.e., one that does not end in infinitely many zeros. Any decimal fraction which does terminate, such as .56, could be replaced with a representation ending in infinitely many 9s. For example, $.56 = .5599999\ldots$. Similarly, the number 1 can be written as $.99999\ldots$. Using the convention that we write all decimal fractions in non-terminating form, we construct the bijection as follows.

Let $r \in (0, 1]$. Being non-terminating, it is of the form $r = .r_1 r_2 r_3 \ldots$ and contains an infinite number of nonzero digits. Split the decimal representation of r into sections where each section consists of either one digit (if the digit is nonzero) or several digits beginning with zero and ending with the next nonzero digit.

For example, say $r = .53070034500038\ldots$. Then split r as shown in Figure 4.16.

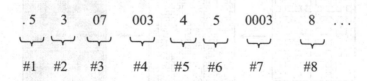

Figure 4.16. Splitting r to form (x, y).

Now detach the odd sections and concatenate them to create a decimal fraction x. Here $x = .50740003\ldots$. Similarly detach the even sections and create the decimal fraction y. In this case $y = .300358\ldots$. Our bijection will be such that the real number $r = .53070034500038\ldots$ is mapped to the point $(x, y) = (.50740003\ldots, .300358\ldots)$ and we define our bijection $r \mapsto (x, y)$ as in Figure 4.16.

In a similar manner we could show there exists a bijection from $(0, 1]$ into the three-dimensional unit cube $\{(x, y, z) : 0 < x \leq 1, \ 0 < y \leq 1, \ 0 < z \leq 1\}$. Concatenate sections numbered $1, 4, 7, \ldots$ to create the decimal fraction x. Concatenate sections numbered $2, 5, 8, \ldots$ to create y. Finally, concatenate sections numbered $3, 6, 9, \ldots$ to form z. The desired bijection is $r \mapsto (x, y, z)$ In other words, dimension has nothing to do with the cardinality of a continuous point set. All such sets have the cardinality of the continuum and it is possible to place them in one-to-one correspondence with each other.

The bijection described previously is not continuous and is therefore not a *stretch* as one would think of stretching a rubber band. Can a line segment be stretched (continuously) in such a way that it fills an area? More generally, is there such a thing as a *space filling curve*? In 1890, Italian mathematician Giuseppe Peano (1858–1932) was the first to show that this was indeed possible. David Hilbert and E. H.

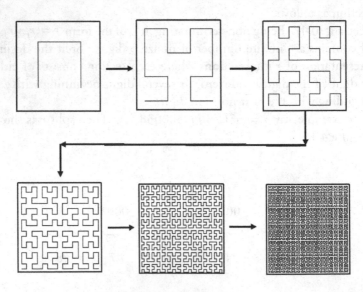

Figure 4.17. Hilbert Curve.

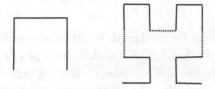

Figure 4.18. Creating the Hilbert Curve.

Moore (1862–1943) followed with their own examples of such space filling curves. Figure 4.17 shows how a one-dimensional curve can fill a two-dimensional square.

The Hilbert Curve is created by starting with the basic shape (a staple) depicted in the upper left. The second step is achieved by replacing the staple of the first step with four half size copies of itself joined by three line segments (Figure 4.18).

The third iteration is achieved as before by replacing each of the four small staples of the second step by four smaller staples and their connecting line segments. The process continues forever and clearly no curve so obtained completely fills the square. It is the limiting curve which fills the square, and this curve can be shown to be continuous. Such space filling curves are not bijections in that they must intersect themselves and therefore are not one-to-one. Nevertheless it is a remarkable stretch!

Can three-dimensional space be filled by a curve? Why not! (See Figure 4.19.)

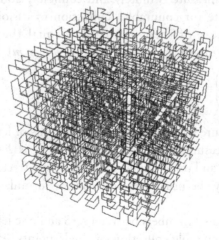

Figure 4.19. Three-dimensional Hilbert Curve.

Cantor Dust

A piece of string one foot in length is cut in half and one of the two pieces is discarded. What's left? We are left with a piece of string one-half foot long. More precisely, if we remove a line segment of length b from a longer line segment of length a, what remains is a segment of length $a-b$. On the other hand, the cardinality of the remaining segment (considered as a point set) is the same as that of the original segment.

$$\frac{1}{2}c = c$$

Cardinality, be it finite or infinite, measures set content and not extent. For example, the intervals $[0, 1]$, $[0, 2]$, and $(-\infty, \infty)$ have the same cardinality (c) and therefore are considered to have the same number of points. Yet the first is one unit in length, the second is two units in length, and the third is of infinite length. Cardinality does not reflect these lengths, and we need some scheme to measure the extent of a point set.

The *Lebesgue measure*, or simply *measure* of a point set, measures the extent or size of the set. For the general n-dimensional point set A, the Lebesgue measure $m(A)$ satisfies the following conditions:

1. The n-dimensional unit cube has measure one.

2. Congruent sets have the same measure. That is, m is invariant under an isometry.

3. Lebesgue measure is finitely and countably additive. That is, for a finite (or countable) collection of disjoint point sets, the measure of their union is to equal the sum of their respective measures. Formally, $m(A \cup B) = m(A) + m(B)$ and

$$m\left(\bigcup_{i=1}^{\infty} A_i\right) = \sum_{i=1}^{\infty} m(A_i).$$

Lebesgue measure satisfies the common notions of length, area, and volume. The measure of the line segment $[a, b]$ is $b - a$. Measures of common two- and three-dimensional point sets (rectangles, spheres, etc.) are given by the usual area and volume formulas. A point has measure zero.

The set of rational numbers between zero and one has measure zero because it is a countable collection of single points, each of measure zero. It follows that the set of irrational numbers between zero and one must be of measure one.

Let's get back to our piece of string and ask the following. If we remove several pieces having a total length of one foot from the original one foot long piece, what's left? Whatever is left, if anything at all, will have a total measure, or length, of zero. But what about the cardinality of the remaining point set? Would it be zero? Could it be anything but zero? If the pieces are removed in a special way, the remaining set, known as the *Cantor set*, or *Cantor dust*, turns out to be quite remarkable.

Begin with a line segment of unit length and remove the middle third. That is, from [0, 1] remove all points strictly between $\frac{1}{3}$ and $\frac{2}{3}$. What remains is $[0, 1] - (\frac{1}{3}, \frac{2}{3}) = [0, \frac{1}{3}] \cup [\frac{2}{3}, 1]$. Now remove the middle third from each of the two remaining pieces. Continue the process through infinitely many steps by always removing the middle third of the remaining pieces. Figure 4.20 shows the first few steps of this process.

It is difficult to illustrate the Cantor set beyond this point; however, it appears as if there would be little if any left of our original segment after infinitely many such steps. Supporting this is the fact that the total length of all intervals removed is one, the length of our original segment. We show this by evaluating the infinite geometric series

$$\frac{1}{3} + \frac{2}{9} + \frac{4}{27} + \frac{8}{81} + \ldots = \frac{1}{3} \sum_{i=0}^{\infty} \left(\frac{2}{3}\right)^i = \frac{1}{3}\left(\frac{1}{1 - \frac{2}{3}}\right) = 1.$$

It follows that the measure, or total length of the remaining point set, is zero. In fact, if we randomly choose a point from the original

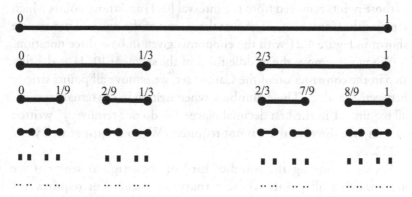

Figure 4.20. Formation of the Cantor set.

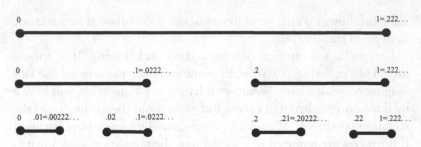

Figure 4.21. Formation of the Cantor set—ternary representation.

unit segment, the probability that it belongs to the Cantor set is zero. So there's nothing left, right? Where's the paradox? In fact, there is something left and much more than we might anticipate. At a minimum the endpoints of the segments on Figure 4.20 remain. So the Cantor set includes

$$0, 1, \frac{1}{3}, \frac{2}{3}, \frac{1}{9}, \frac{2}{9}, \frac{7}{9}, \frac{8}{9}, \frac{1}{27}, \ldots .$$

At this stage we might suspect that the Cantor set will consist of a countable number of points whose total measure is zero. But the truth is that once we remove all of the intervals, what remains is an uncountable set of points, no less than the number of points with which we started! That is, the set known as Cantor dust, of measure zero, can be put into one-to-one correspondence with the points of the original and complete unit line segment. It's as if we could somehow rearrange the points of the Cantor set to reconstruct the original interval.

To see why this is so, consider the base three (ternary) representation of those points removed from the interval [0, 1] and those points which remain (the Cantor set). The first three rows of the removal process are shown in Figure 4.21 with the endpoints given in base three notation.

When we remove the middle third of the segment [0, 1] at the first step in the construction of the Cantor set, we remove all points strictly between $\frac{1}{3}$ and $\frac{2}{3}$. These numbers, when written in the ternary system, all require a 1 in the first decimal place. We do not remove $\frac{1}{3}$, written as .1 in base three, as the 1 is not required. We could just as well write .0222. . . .

So, by removing the middle third of the original segment we are removing all points whose ternary representation requires a 1 immediately following the decimal point.

The second step in the construction of the Cantor set removes $(\frac{1}{9}, \frac{2}{9})$ from $[0, \frac{1}{3}]$ and $(\frac{7}{9}, \frac{8}{9})$ from $[\frac{2}{3}, 1]$. This removes all points whose ternary representation requires a 1 in the second decimal place. And the third stage removes all points requiring a 1 in the third decimal place, and so on. We conclude that the Cantor set (that which remains after all such intervals are removed) consists of those points which can be represented entirely by 0s and 2s following the decimal point.

But how do we know the Cantor set contains uncountably many such points? There is an obvious mapping of all numbers written with 0s and 2s after the decimal point to numbers written with 0s and 1s after the decimal point. Simply change each 2 to a 1. And how many numbers can be written using 0s and 1s after the decimal point? There are uncountably many such numbers since the binary representation of any number between zero and one involves only 0s and 1s. Our conclusion is that the Cantor set (of measure zero) is of cardinality c and can be mapped continuously, or stretched, onto the closed interval $[0, 1]$ (of measure one). A little goes a long way if we do it right!

So the Cantor set must contain more points than just the endpoints of the remaining line segments. For example, the number $\frac{1}{4}$ (base ten representation .25) has a ternary representation .020202... and so belongs to the Cantor set. Likewise $\frac{3}{4}$ has a ternary representation .202020... and $\frac{1}{13}$ has a ternary representation .002002002.... Each belongs to the set. In actuality, the endpoints constitute an infinitesimally small part of the Cantor set.

A graphical way to confirm the cardinality of the Cantor set is to characterize each point of the set by the path required to reach the point, starting at the top of Figure 4.22. An infinite sequence of left-right decisions, or branches, are required to arrive at the target point of the Cantor set. The sequence illustrated, LRRLR..., defines a specific point of the set. If we replace L with 0 and R with 1, then we have infinite sequences of zeros and ones specifying points of the Cantor set. Such sequences, preceded by the decimal point, represent real numbers between 0 and 1, a set with cardinality c. This confirms the fact that there are as many points of the Cantor set as there are points on the original line segment $[0, 1]$.

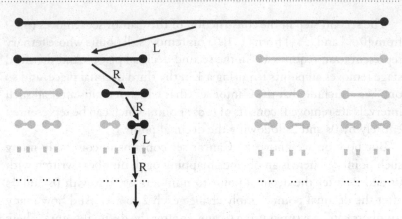

Figure 4.22. Binary branching to specify a point in the Cantor set.

To put all of this in a more extreme way, consider the following. The particles of Cantor dust (cardinality c) reside in one-dimensional space and are of measure zero. They can be continuously repositioned (stretched) to cover the unit interval [0, 1]. The points of this interval can be continuously repositioned to create a space filling curve, filling the unit cube. The points of the cube can be continuously repositioned to coincide with the points of a cube of any finite volume.

We can conclude that the particles of Cantor dust, of measure zero, can be continuously repositioned to fill the volume of any cube, no matter how large!

A similar construction can be done on the unit square to form a two-dimensional analog of the Cantor set. Begin by removing the center one-ninth of a unit square. Then remove the centers of the eight smaller squares, and so on. The remaining set of points, known as the *Sierpiński Carpet* (Figure 4.23), is of measure zero since the total area removed is given by

$$\frac{1}{9} + \frac{8}{81} + \frac{64}{729} + \dots = \frac{1}{9} \sum_{i=0}^{\infty} \left(\frac{8}{9} \right)^i = \frac{1}{9} \left(\frac{1}{1 - \frac{8}{9}} \right) = 1$$

Yet, as before, the cardinality of its point set is c and the points of the carpet can be put in one-to-one correspondence with each and every point of the original unit square.

In three dimensions the construction creates the Sierpiński-Menger Sponge (Figure 4.24).

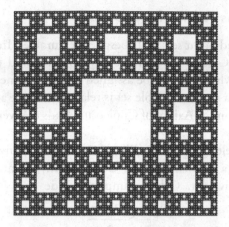

Figure 4.23. The Sierpiński Carpet.

The Cantor sets (in one, two, or three dimensions) all exhibit the fractal quality of being self-similar. Any piece of the set, no matter how small, can be magnified to appear the same as the original. These sets may be the oldest mathematically contrived fractals.

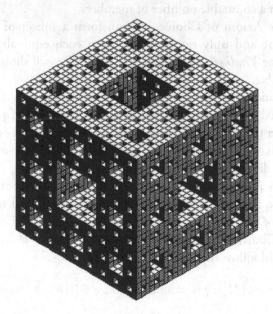

Figure 4.24. The Sierpiński-Menger Sponge.

The Vitali Construction Paradoxes

Must all bounded point sets be Lebesgue measurable? In 1905, Italian mathematician Giuseppe Vitali (1875–1932) showed that bounded point sets exist which can not be assigned a Lebesgue measure. Vitali's construction of a nonmeasurable set is relatively simple and we present it here. It relies on the Axiom of Choice and leads to several remarkable Baby BTs.

Let a and b belong to $[0, 1]$. We say a and b are equivalent (writing $a \sim b$) if $a - b$ is a rational number. Consequently, all rational numbers in $[0, 1]$ are equivalent. Other examples include

$$\frac{1}{\pi} \sim \frac{2 + \pi}{2\pi}$$

because these two numbers differ by the rational number $\frac{1}{2}$. The numbers $\frac{1}{2}$ and $\frac{1}{\sqrt{2}}$ are not equivalent because their difference is irrational. We can now partition $[0, 1]$ into equivalence classes where each class consists of a set of numbers which are equivalent. That is, two numbers chosen from a given class differ by a rational number and two numbers chosen from respectively different classes differ by an irrational number. The interval $[0, 1]$ consists of uncountably many classes, each consisting of a countable number of members.

Using the Axiom of Choice, we can form a subset of $[0, 1]$ by selecting one and only one number from each equivalence class. Known as the *Vitali set*, we designate it as M. We will show that M is nonmeasurable.

For each rational number q let $M_q = M + q = \{x + q : x \in M\}$. That is, M_q is M shifted q units to the right if q is positive and to the left if q is negative. We call M_q a translate of M. Note $\cup \{M_q : q \text{ is rational}\} = (-\infty, \infty) = \mathbb{R}$. That is, we have partitioned the set of all real numbers into a countable collection of disjoint congruent sets.

We claim M is nonmeasurable and show this by contradiction. First assume M is measurable. Then for each rational number q, $m(M_q) = m(M)$. If $m(M) = 0$ then $m(\mathbb{R}) = 0$ because $\mathbb{R} = \cup \{M_q : q \text{ is rational}\}$ Clearly this is impossible. But $m(M) > 0$ is also impossible since it would follow that

$$m([0, 2]) \geq m(\cup \{M_q : q \text{ is rational and } 0 \leq q \leq 1\}) = \sum_{\substack{0 \leq q \leq 1 \\ q \text{ rational}}} m(M_q) = \infty.$$

This too is impossible as $m([0, 2])=2$. Therefore M (and each M_q) is not Lebesgue measurable.

Some might find it disagreeable that such sets exist; but, such is mathematics! Nonmeasurable sets play a fundamental role in the Banach-Tarski Theorem in that the sphere will be decomposed into subsets, some of which are nonmeasurable. Here we give some less dramatic paradoxes involving the Vitali construction.

> *A subset of $[0,2]$ can be partitioned in such a way that the pieces can be translated so that their union is the entire real number line.*

This is quite amazing in that there is no stretching (magnification) as discussed earlier in this chapter. The pieces are isometrically (distance preserving) translated so as to make up the entire real number line. (The resolution of this oddity is similar to the resolution of the Banach-Tarski Theorem itself. The discussion is given in Chapter 6.)

To show this, take $\cup\{M_q : q$ is rational and $q \in [0, 1]\}$ as the partition of a subset of $[0, 2]$. Since the set of rationals belonging to $[0, 1]$ is a countable set and the set of all rational numbers is also a countable set, we are able to put the rationals in $[0, 1]$ in one-to-one correspondence with all rational numbers. Let this correspondence be given by the function $f(q) = r$ where q is a rational number in $[0, 1]$ and r is an unrestricted rational number. For each q, translate M_q to $M_{f(q)}=M_r$. Note $\cup\{M_r : r$ is rational$\} = (-\infty, \infty) = \mathbb{R}$. So, the subsets of the original partition have been respectively translated so as to form the entire real number line.

An equally remarkable result can be found by constructing the Vitali set on the circumference of a circle. We show the following:

> *A circle of any given radius can be partitioned into two sets in such a way that each can be decomposed and reassembled to form the complete original circle.*

Two circles from one! It sounds as if we're getting close to the Banach-Tarski Theorem.

We form the Vitali set on the circle by calling two points equivalent if one can be obtained by the other via a rotation by a rational multiple of one revolution. (We take this rational number to be between zero and one.) This is analogous to the previous Vitali set where two points were considered equivalent if the distance between the points was

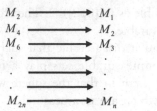

Figure 4.25. Rotating the even indexed subsets to form the complete circle.

rational. As before we partition the set into equivalence classes and use the Axiom of Choice to define a set M consisting of exactly one point from each class. Because the rationals between zero and one form a countable set, we can designate the rotations as $\rho_1, \rho_2, \rho_3, \dots$. Then let $M_i = \rho_i M$ for $i = 1, 2, 3, \dots$.

The complete circle is given by $\cup\{M_i : i = 1, 2, 3, \dots\}$. Since all M_i are congruent, we can individually rotate the even indexed M_i to coincide with all the M_i as in Figure 4.25.

Thus, the even indexed M_i can be individually rotated to form the complete circle. In a similar manner we could individually rotate the odd indexed M_i to form the complete circle (Figure 4.26).

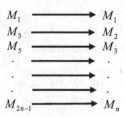

Figure 4.26. Rotating the odd indexed subsets to form the complete circle.

Summarizing, we have partitioned the original circle into two sets in such a way that each is equidecomposable to the original circle. Two circles from one! This construction truly deserves being described as a Baby BT.

We could use a similar argument to create any number of circles from the single original circle. For example, to form three circles from the original, define the M and M_i as before and then rotate the M_i as shown in Figure 4.27.

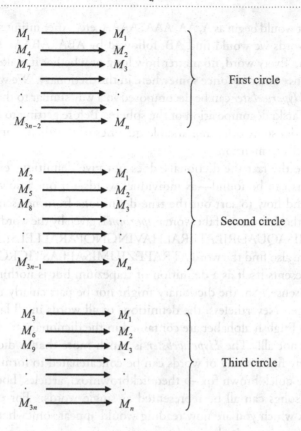

Figure 4.27. Three circles from one.

Stewart's Hyperwebster Dictionary

Ian Stewart, professor of mathematics at the University of Warwick and one of the world's best known authors of popular mathematics, gives a wonderful example of paradoxical decomposition involving little, if any, mathematics. Yet, the decomposition is in many respects similar to the decomposition used in the Banach-Tarski Theorem. In *From Here to Infinity*, Stewart discusses [Stewart 96, pp. 175–176] the ultimate dictionary—the *Hyperwebster*—which lists all possible words formed using the twenty-six letters of the English alphabet. The words, each consisting of a finite number of letters, are listed in alphabetical order. The dictionary is simply a list of words, whether or not they make sense, with no definitions explicitly given.

The list would begin as A, AA, AAA, AAAA, etc. After infinitely many of such words we would find AB, followed by ABA, ABAA, ABAAA, and so on. Every word, no matter how long or whether it makes sense, appears once and only once somewhere in the dictionary. We will show how the *Hyperwebster* can be decomposed in a way similar to that of the Banach-Tarski decomposition of the sphere. Before getting to that, we first consider some other remarkable qualities of this dictionary which Stewart fails to mention.

Despite the fact the dictionary does not give definitions explicitly, definitions can be found—as individual words—if one knows where to look and how to sort out the true definitions from nonsense. For example, the definition of the word *trapezium* is given by the word ATRA-PEZIUMISAQUADRILATERALHAVINGNOPARALLELSIDES. Of course, we also find the word ATRAPEZIUMISAPLASTICRODENT which presents itself as a definition of trapezium but is nothing more than nonsense. So, the dictionary might not be particularly useful as a reference. Nevertheless, the definitions of all words in all languages using the English alphabet are contained in the dictionary.

That's not all! The *Hyperwebster* is much more than a dictionary. In that any finite string of words can be concatenated to form a single word (the quick brown fox → thequickbrownfox), articles, books, and lyrics to songs can all be represented as single words. For example, this book which you are now reading would appear somewhere in the *Hyperwebster* as a single word consisting of approximately 300,000 letters. Every published (and unpublished) work of any sort using the English alphabet would appear as a single word. Potential cures to diseases which presently have no known cure would appear as words. Bogus treatments would appear as well and a staff of medical experts would be required to sort out the good from the garbage. We might as well think of the *Hyperwebster* as a universal library containing all that has ever or will ever be written using the English alphabet. It sounds like a must for anyone's personal library!

Assume the hypothetical publisher, Above and Beyond, Ltd., decides to publish the *Hyperwebster* as a 26 volume set. All words beginning with the letter A are contained in Volume A, those words beginning with B are contained in Volume B, and so on. Words are listed in alphabetical order within their respective volumes (Figure 4.28).

Volume A: A, AA, AAA, . . . , AB, ABA, ABAA, . . . , ABB, . . .

Volume B: B, BA, BAA, . . . , BB, BBA, BBAA, . . . , BBB, . . .

Volume C: C, CA, CAA, . . . , CB, CBA, CBAA, . . . , CBB, . . .

Volume Z: Z, ZA, ZAA, . . . , ZB, ZBA, ZBAA, . . . , ZBB, . . .

Figure 4.28. The twenty-six volume *Hyperwebster*.

As press time draws near, A&B Ltd. begins to realize the enormity of its task and decides to cut some corners. For Volume A, it is decided to omit the first letter (A) of each word as the reader surely will know, by the volume title, these are the A words. (This is similar to the common practice of leaving off the decimal point in tables of numbers each of which begins with the decimal point.) The same will be done for Volume B by eliminating the first letter (B) of each word listed in the volume. In similar fashion it is decided to leave off the first letter of all words given in the entire *Hyperwebster*. The reader will have no problem mentally appending the title letter of the volume to the front of each word contained in that volume. A&B figures to save a lot of money by doing this, at little inconvenience to the reader.

Just hours before going to print, another cost cutting discovery is made. A careful inspection of the 26 volumes reveals that they are identical, except for their covers. Look at Figure 4.28 and compare the contents of each volume once the first letter of every word is removed. The contents are identical! With the exception of cover art, the *Hyperwebster* has just been decomposed into identical 26 copies of itself! Why print 26 identical volumes as part of a 26 volume set? It is decided to print only the first volume (Volume A) and change the cover title from "Volume A" to "*Hyperwebster*."

A&B takes a last look at the single volume before going to press. They decide to divide the single volume into chapters with Chapter A containing all words beginning with the letter A, Chapter B containing all words beginning with the letter B, and so on. As before, it is decided to leave off the first letter of each word within any given chapter as the reader will know, by the chapter's title, what that letter should be. As press time draws near, the staff at A&B is embarrassed to note that each chapter contains the exact same content. The reason is the same as for the 26 identical volumes. So, once again, A&B has to rethink the project and finally decides to only publish Chapter A. The title, of course, is changed from "Chapter A" to "*Hyperwebster.*"

Well you can guess what happens next. The staff at A&B decides to divide the chapter into 26 sections—Section A, Section B, and so on. Every time they are ready to go to press, it is discovered that . . .

Summarizing, the *Hyperwebster*, beyond the wonder of its universal content, possesses the additional property of being decomposable into infinitely many copies of itself, each of which is decomposable into infinitely many copies of itself, etc.

And for infinitely many reasons, A&B cuts its losses and abandons the project.

Stewart notes [Stewart 96, p. 176], "In spirit, the Banach-Tarski Paradox is just the *Hyperwebster* Paradox wrapped round a sphere." The *Hyperwebster* decompositions produce copies of identical content while the duplication version of the Banach-Tarski Theorem produces solids of identical size and shape.

The Sierpiński-Mazurkiewicz Paradox

At the beginning of this chapter when discussing shifting to infinity, we saw how it would be possible to decompose a point set into two subsets in such a way that one of the subsets was congruent to the original set. For example, the point set $\{1, 2, 3, \ldots\}$ can be decomposed into the two subsets $\{1, 2\}$ and $\{3, 4, 5, \ldots\}$ with the latter being congruent to the original by shifting the latter (from infinity) two units to the left. The set $\{1, 2\}$ is irrelevant to this discussion and we disregard it.

Polish mathematician Wacław Sierpiński (1882–1969) asked if it would be possible to decompose a set E into two disjoint subsets E_1 and E_2 in such a way that each subset would be congruent to the original set E. That is, given E, do the subsets E_1 and E_2 exist where $E = E_1 \cup E_2$, $E_1 \cap E_2 = \emptyset$, $E_1 \cong E_2 \cong E$?

Sierpiński's student Stefan Mazurkiewicz (1888–1945) answered in the affirmative by defining such a set in the plane. The result, known as the *Sierpiński-Mazurkiewicz Paradox*, is arrived at without using the Axiom of Choice and has the $1 = 2$ flavor of the Banach-Tarski Paradox.

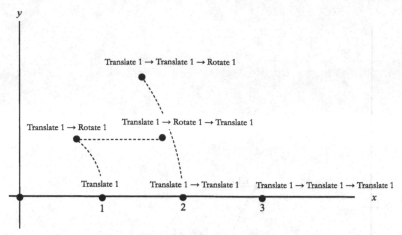

Figure 4.29. Some points of the set E.

(It is worth noting that Sierpiński and Mazurkiewicz were the co-founders of *Fundamenta Mathematicae*, the journal in which the Banach-Tarski Theorem was published.)

Let E consist of the origin O and all points obtainable from O by a finite combination of the following two operations:

1. Translate right one unit.

2. Rotate counterclockwise one radian ($\frac{180}{\pi}$ degrees) about O.

E is a countable set of points representing an infinitesimally small part of the entire plane. We show in this chapter's appendix that every point of E is uniquely represented by a combination of translations and rotations as described above.

Figure 4.29 shows some points of E.

We now partition E into the desired subsets E_1 and E_2 by defining E_1 to be the set of points obtained by shifting E to the right one unit and E_2 to be the set of points obtained by rotating E one radian counterclockwise. Clearly, $E \cong E_1$, $E \cong E_2$, and $E = E_1 \cup E_2$. The fact that E_1 and E_2 are disjoint (having no common points) is discussed in this chapter's appendix. We now have the desired partition.

Another way of looking at this paradox is to consider each point of E as being created by a finite sequence of translations and rotations. If we let T denote the translation and R the rotation, then, following the convention given in Chapter 3, we could describe each point of E by

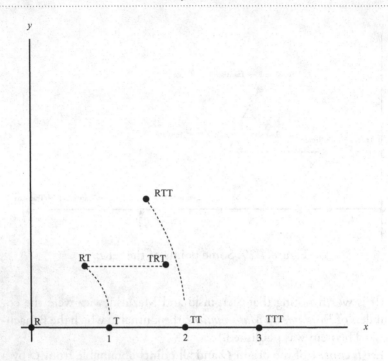

Figure 4.30. Points of the set E as *words*.

a finite string of letters with the transformations given in order from right to left.

That is, the first transformation will be the rightmost character and the last transformation will be the leftmost character. So, a translation followed by a rotation would be denoted by the "word" RT and three translations followed by two rotations would be given by RRTTT. Every point of E is associated with such a word. (We describe the origin O by the word R since rotating the origin yields itself. To avoid further confusion, this will be the only word whose rightmost character is the letter R. For example, TRR is best described as T.)

Figure 4.30 uses this convention to identify the points of the previous figure.

Now we could think of E_1 as all points, or *words*, whose leftmost character is T. Similarly E_2 is described using all words whose leftmost character is R. Remember that it is our convention not to list words whose rightmost character is R (with the exception of the origin—denoted as R) to avoid multiple representations for each point.

So, sets E_1 and E_2 could be described as follows, if the words of each set are given in order of increasing length:

E_1: T, TT, TRT, TTT, TRRT, TRTT, TTRT, TTTT, . . .

E_2: R, RT, RRT, RTT, RRRT, RRTT, RTRT, RTTT, . . .

Does this look familiar? We could describe it, for lack of a better description, as the *Sierpiński-Mazurkiewicz Hyperwebster*. For an alphabet consisting of only the two letters R and T, we could divide all words into two sets, those words beginning with a leftmost character of T and those words beginning with a leftmost character of R. Removing the leftmost character of all words in any given set yields all words of the original set E. (Removing the leftmost character of T yields R since we could think of T as being equivalent to TR. Similarly, removing the leftmost character of R yields R since we could think of R as being equivalent to RR.)

One wonders if a *bounded* subset of the plane (E is unbounded) can be partitioned into disjoint subsets, each of which is congruent to the original. It is known that it can not be done using only two pieces [Wagon 85, p. 196]. We see in Chapter 5, as part of the proof of the Banach-Tarski Theorem, that there exist bounded sets of points in three-dimensional space with the above paradoxical property.

Appendix

The Sierpiński-Mazurkiewicz Paradox was illustrated geometrically, as points in the plane, and shown similar to Stewart's *Hyperwebster*. The point set E can be described algebraically by allowing points in the plane to represent complex numbers (in the complex plane). Readers familiar with the trigonometric and exponential form of complex numbers will appreciate the following analysis.

Every complex number $z = a + bi$, $a \in \mathbb{R}$, $b \in \mathbb{R}$, $i^2 = -1$, can be written in trigonometric form as $z = r(\cos\theta + i\sin\theta)$. In exponential notation we write $z = r e^{i\theta}$. Translating one unit to the right is equivalent to adding 1 to the number. Rotating counterclockwise one radian is equivalent to multiplying by e^i. For example, the points obtained by translating the origin O two units to the right followed by a rotation of one radian is given by $0 \to 1 \to 2 \to 2e^i$. Rotating this point one additional radian takes us to $2e^{2i}$. Shifting this point to the right one unit yields $1 + 2e^{2i}$. It follows that every point of E can be represented as a polynomial in e^i of the form $z = a_0 + a_1 e^i + a_2 e^{2i} + \ldots + a_n e^{ni}$ where the a_i are nonnegative integers and n is a positive integer. The set E_1 consists of points with algebraic representations such that $a_0 > 0$. The set E_2 consists of points with representations such that $a_0 = 0$. Shifting any point (z) to the right one unit adds 1 to a_0. Rotating any point (z) one radian counterclockwise increases the coefficient of i in all exponents by one unit.

The properties of E given in this chapter should be apparent from the algebraic description of E. Clearly $E = E_1 \cup E_2$. It now becomes clear why $E_1 \cap E_2 = \emptyset$. If it were the case that a point of E had two distinct polynomial representations (two distinct words in letters T and R) then e^i would be the solution to some polynomial equation. This is impossible as it has been shown that e^i is transcendental (satisfies no polynomial equation with rational coefficients).

5 Statement and Proof of the Theorem

Infinity is where things happen that don't.
—An anonymous schoolboy

Both statement and proof of the theorem of Stefan Banach and Alfred Tarski are straightforward. It is only when one contemplates the conclusion of the theorem that the magic appears. This chapter presents the proof of the theorem in as much detail as possible, without getting bogged down in specific mathematical technicalities which are better left to be investigated outside of this presentation. Mathematical notation is kept to a minimum; however it is assumed that the reader is familiar with the contents of Chapter 3. The interpretation and resolution of this theorem's stunning conclusion are presented in Chapter 6.

Statement of the Banach-Tarski Theorem

A solid ball may be separated into a finite number of pieces and reassembled in such a way as to create two solid balls, each identical in shape and volume to the original.

Formally, this duplication version of the theorem can be stated as follows:

The unit ball $B=\{(x,y,z) : x^2 + y^2 + z^2 \le 1\}$ can be partitioned into two sets B_1 and B_2 such that $B \sim B_1$ and $B \sim B_2$. Here "$\sim$" means "is piecewise congruent to" or "is equidecomposable to."

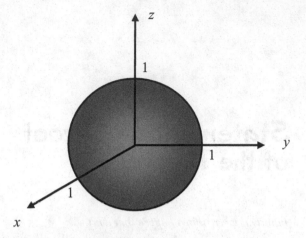

Figure 5.1. The unit ball B.

A different, but equally paradoxical version of the theorem asserts that a solid of any shape and volume can be decomposed and reassembled to form another solid of any specified shape and volume. Thus, the theorem is sometimes referred to as the *pea and the sun* paradox. It will be shown that this magnification version follows from the duplication version.

Proof of the Banach-Tarski Theorem

Without loss of generality, we focus on a ball of radius one (the unit ball) centered at the origin of the standard xyz Cartesian (rectangular) coordinate system. We can think of the ball as a set of points $B=\{(x,y,z): x^2 + y^2 + z^2 \leq 1\}$ (Figure 5.1).

We will show that B can be partitioned into a finite number of sets and reassembled in such a way as to form two copies of itself—B_1 and B_2.

The proof is presented in three steps. Step I creates a group of rotations of the unit sphere (surface of the ball). Step II uses these rotations to partition the sphere into subsets possessing a remarkable property known as the *Hausdorff Paradox*. Lastly, Step III extends the above mentioned property from the spherical surface to the solid ball. This will conclude the proof of the duplication version of the Banach-Tarski Theorem.

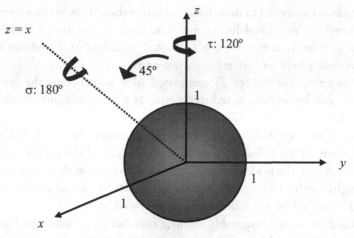

Figure 5.2. Basic rotations τ and σ.

Step I—The Group of Rotations *G*

The *rotation* of a figure (a set of points) about an axis in three-dimensional space is a rigid motion such that each point of the figure moves in a circular path about the axis in a plane perpendicular to the axis. Having said this, we may just assume that a rotation is precisely what we would expect it to be—a turning of the figure in such a way that there is no distortion of the figure.

We begin by defining two basic rotations of the unit sphere. Let τ denote a clockwise rotation of 120° about the *z* axis and let σ denote a clockwise rotation of 180° about the line $z = x$ in the *xz* plane. This line passes through the origin and makes an angle of 45° with the *z* axis (Figure 5.2).

These two basic rotations can be combined sequentially in a countably infinite number of ways to yield infinitely many other rotations. For example, two τ rotations, denoted as ττ or $τ^2$, corresponds to a rotation of 240° about the *z* axis. The rotation $σ^2$ would represent a rotation of 360° about the line $z = x$ and the sphere would be rotated all the way around to its initial position. We call such a rotation the identity rotation and denote it by *I*. So it follows that $τ^3 = σ^2 = I$.

The single rotation equivalent to τ followed by σ is στ. Note we use the previously adopted convention of writing the rotations from right to left with the first rotation being listed on the far right and following rotations being listed, in order, from right to left.

Rotations composed of these basic rotations should always be written in reduced, or simplified form. For example, the rotation τ^7 is best written as τ because $\tau^7 = \tau^3\tau^3\tau = I I \tau = \tau$. Similarly $\tau^4\sigma^3\tau$ reduces to $\tau\sigma\tau$ because $\tau^4\sigma^3\tau = \tau^3\tau\sigma^2\sigma\tau = I\tau I\sigma\tau = \tau\sigma\tau$.

So, any rotation (except I) composed of a sequence of the basic rotations can be written, in reduced form, as a string of symbols, each of the form τ, τ^2, or σ. We define the *length* of a rotation to be the number of such symbols used to define the rotation. So, the length of the rotation $\sigma\tau^2\sigma\tau$ is four. (We define the length of the rotation I to be zero.) Again, the reader is reminded that these strings should be read from right to left. That is, the single rotation $\tau\sigma\tau^2\sigma$ should be thought of as σ, followed by τ^2, followed by σ, followed by τ.

In total, there is a countably infinite number of possible rotations formed as above and we refer to the collection of all such rotations as the group of rotations G.

The following theorem is key to our discussion. Its proof is given in this chapter's appendix.

The Uniqueness Theorem

Every rotation in G has a unique, reduced form representation.
That is, if two reduced form rotations appear different, they do,
in fact, represent different physical rotations.

Having defined this group of rotations G, we now proceed to partition G into three subsets—G_1, G_2, and G_3. We will specify a rule which assigns each rotation of G to one and only one of the specified subsets. Think of the assignment as a sorting process, placing the rotations, one at a time, into their correct subsets. The assignments are made in order of the lengths of the rotations, beginning with the identity rotation I, then the rotations of length one, two, etc. Of course, it is an infinite process with each member of G ultimately receiving its assignment to one of the three subsets. It is a recursive process in that a rotation's assignment is based, in part, on a previous rotation's assignment.

The assignment process is efficiently summarized by Figure 5.3. The process begins by assigning the identity rotation I to G_1, τ and σ to G_2, and τ^2 to G_3. At this point, all rotations up to length one have been assigned. As indicated, the assignments for all rotations of length two are made. Then it becomes possible to assign the rotations of length three, four, and so on.

	If $\alpha \in G_1$	If $\alpha \in G_2$	If $\alpha \in G_3$
If the leftmost character of α is τ or τ^2	assign $\sigma\alpha$ to G_2	assign $\sigma\alpha$ to G_1	assign $\sigma\alpha$ to G_1
If the leftmost character of α is σ	assign $\tau\alpha$ to G_2	assign $\tau\alpha$ to G_3	assign $\tau\alpha$ to G_1
	assign $\tau^2\alpha$ to G_3	assign $\tau^2\alpha$ to G_1	assign $\tau^2\alpha$ to G_2

Figure 5.3. Partitioning G into G_1, G_2, and G_3.

For example, we have been given the assignments of all length one rotations and now wish to assign the length two rotations to their respective subsets. The rotation $\tau\sigma$ would be assigned to G_3 because σ (having a leftmost character of σ) has already been assigned to G_2 and attaching τ in front obligates us, by the scheme, to place $\tau\sigma$ in G_3. To assign, say, $\sigma\tau^2$, we note that τ^2 (having a leftmost character of τ^2) has already been assigned to G_3. Attaching σ at the left requires we assign $\sigma\tau^2$ to G_1. When we get around to assigning rotations of length three and wish to assign $\tau\sigma\tau^2$ to its subset, we would choose to assign it to G_2 because $\sigma\tau^2$ (having a leftmost character of σ) has already been assigned to G_1 and attaching τ up front requires us to place $\tau\sigma\tau^2$ in G_2.

A few members of the subsets G_1, G_2, and G_3 are shown in Figure 5.4.

Note that as a consequence of the Uniqueness Theorem, the process continues ad infinitum listing all rotations in G, each of which is physically unique.

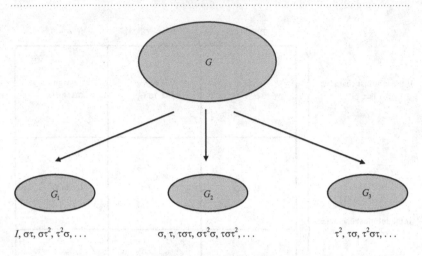

$I, \sigma\tau, \sigma\tau^2, \tau^2\sigma, \ldots$ $\qquad\qquad$ $\sigma, \tau, \tau\sigma\tau, \sigma\tau^2\sigma, \tau\sigma\tau^2, \ldots$ $\qquad\qquad$ $\tau^2, \tau\sigma, \tau^2\sigma\tau, \ldots$

Figure 5.4. Members of the subsets.

A clever way of envisioning the creation and sorting of the rotations was devised by Robert French, currently at the University of Liege, Belgium. He created a Rube Goldberg type machine [French 88, p. 27] which constructs the rotations and sorts them into their respective sets G_1, G_2, or G_3. The schematic shows the device to consist of three hoppers, three appenders, three sorters, and three collection bags named G_1, G_2, and G_3 (Figure 5.5).

The hoppers are to receive copies of previously created and sorted rotations, sent to them by the copiers. Each copied rotation is then passed on to the appender which increases the length of the rotation by one, in all possible ways, by appending σ, τ, or τ^2 at the left end. For example, the rotation $\sigma\tau$ would be appended (on the left) yielding $\tau\sigma\tau$ and $\tau^2\sigma\tau$. There is no other way to increase the length by one (on the left) and still have the rotation represented in simplified form. The sorter then sends each newly created rotation to a specific copier, based on the leftmost character, as indicated by the schematic. The copier makes a copy of the newly created rotation, sends the copy up to the hopper as shown, and drops the newly created rotation into the collection bag below. The process continues, with all three bags shown at the bottom collecting their rotations.

To fire up the process, we begin by placing the identity rotation I in G_1 and placing a copy of I in the hopper on the far left, above G_1. The

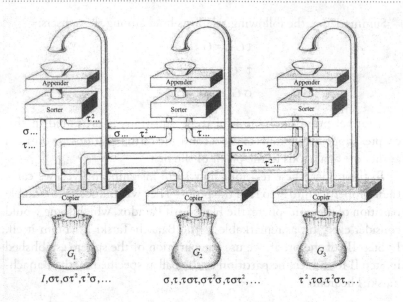

Figure 5.5. Robert French's rotation maker [French 87].
(Reprinted with the permission of Springer-Verlag.)

rest is automatic as the machine does it all. The identity rotation I (of length zero) is appended to form the three rotations of length one—σ, τ, and τ^2. The sorter sends σ and τ to G_2 as shown by the schematic. The rotation τ^2 is sent to G_3. The rotations are copied, and then dropped into the collection bags below. The copies are sent to the hoppers above and the process continues ad infinitum.

The three subsets thus created are related in a peculiar yet significant way. Any given rotation in G_1, if immediately followed by the basic rotation τ (denoted by writing τ immediately to the left of the given rotation), yields a rotation in G_2. For example, the rotation $\sigma\tau$ is in G_1 and the rotation $\tau\sigma\tau$ is in G_2. Furthermore, *each and every* rotation in G_2 is related to a specific rotation in G_1. In fact, we could think of G_2 as being the set of all rotations of G_1, each followed by τ (with τ attached on the left).

Mathematically we write $\tau G_1 = G_2$. Similarly, we can verify $\tau^2 G_1 = G_3$. And finally, it could be verified that following each rotation in G_1 by σ (attaching σ on the left) would give us every rotation in $G_2 \cup G_3$. We write $\sigma G_1 = G_2 \cup G_3$.

Summarizing, the following relations hold among the subsets:

$$\tau G_1 = G_2$$
$$\tau^2 G_1 = G_3$$
$$\sigma G_1 = G_2 \cup G_3$$

A formal proof of the first of these equations is presented in this chapter's appendix. Proofs of the second and third equations are omitted as they are analogous to the proof of the first.

To the reader, these relations may seem insignificant; yet, we carry these properties into Step II of the proof. They will induce a remarkable partition of the unit sphere, the Hausdorff Paradox, which some would consider every bit as remarkable as the Banach-Tarski Theorem itself. In Step III of the proof, we use the partition of the sphere established in Step II to induce the partition of the ball as specified by the Banach-Tarski Theorem.

Step II—Partitioning the Unit Sphere S into Two Copies of Itself

In this step of the proof, we will take the unit sphere, the surface or *skin* of the unit ball,   separate and reassemble it into two sets, each of which is identical in shape to the original unit sphere. To say the least, this is paradoxical and would be considered by some as stunning as the Banach-Tarski Theorem itself. In fact, once we have the unit sphere partitioned in Step II, it becomes a rather small step to extend the process to the ball, as will be done in Step III.

There are two essential mathematical tools we need in Step II. First, we need the three relations among the subsets of G, as previously established. Second, we will need to invoke the Axiom of Choice. The arguments presented in Step II are straightforward and take us through the Hausdorff Paradox to finally arrive at the *two spheres from one* conclusion.

We begin by noting that every rotation in G of the unit sphere has two *poles*. A pole is a point which remains fixed for a given rotation. For example, the rotation τ which rotates the unit sphere 120° about the z axis, has the two poles $(0, 0, 1)$ and $(0, 0, -1)$, much like the north and south poles of the earth. So every rotation in G has its axis and at the ends of the axis we find the two poles associated with the rotation. In that every rotation in G has two poles and there is a countable infinity of rotations in G, we know there are countably many poles on the unit sphere associated with G. Let P denote the set of all poles associated with G on the unit sphere. Let $S-P$ denote all other points on the sphere. Clearly there are infinitely many points in both P and $S-P$. Yet, there are far more points in $S-P$ than P. Recall from Chapter 3 that there are *orders of infinity* and, in a sense, some infinities are larger than others. To put it another way, if a point were randomly chosen from the surface of the sphere, it would almost certainly belong to $S-P$, and not to P. The fractional part of the sphere corresponding to P is infinitesimally small.

Every point in $S-P$ can be thought of as being connected to a countable infinity of other points in $S-P$ via the rotations in G. If any two points in $S-P$ are so connected, we say they belong to the same *orbit*. There is an uncountable infinity of such orbits which make up $S-P$. We now use the Axiom of Choice to select one point from each of these orbits and collectively define these points so chosen as the set C. There is no other way to do this, other than using the Axiom of Choice. These orbits have no quality which allows us to define one member of each. We must simply assume that such a choice is possible and we create the set C by choosing exactly one point from each of these orbits. The set C of points is called the *choice set* as it was formed using the Axiom of Choice.

Summarizing, the surface S consists of a set of poles P, and the remaining points $S-P$. These remaining points, $S-P$, naturally separate into infinitely many orbits. We invoke the Axiom of Choice to create a set C by choosing one point from each of these orbits.

We make the following observations with respect to C:

1. C is uncountably infinite.

2. C and P have no points in common.

3. No point in C can be rotated to any other point in C by one of the rotations in G.

4. If every point in C were to be rotated by every rotation in G, we would ultimately get every point in $S-P$.

Keep in mind that the set $S-P$ represents almost every point on the sphere S, with the exception of the poles P. We will now focus on the set $S-P$, and deal with the poles later.

The Unit Sphere S

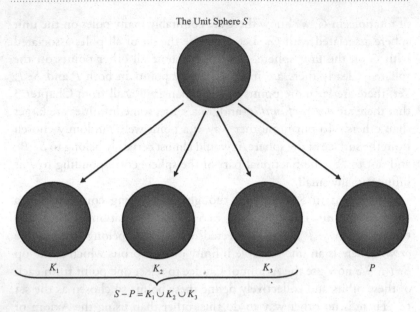

Figure 5.6. The Hausdorff partition of S.

If we apply every rotation of G_1 to C, we will get a set of points on the sphere which we can denote as $G_1 C$. Similarly, if we apply every rotation of G_2 to C, we get another set of points on the sphere which we denote as $G_2 C$. Lastly, we apply the rotations of G_3 to C obtaining $G_3 C$. These three sets of points, $G_1 C$, $G_2 C$, and $G_3 C$ form a partition of $S - P$. That is, these three sets are disjoint (have no points in common) and include all points of $S - P$. With this in mind, we have now partitioned the entire sphere into four disjoint sets—$G_1 C$, $G_2 C$, $G_3 C$, and the set of poles P. To make the notation a bit clearer, let $K_1 = G_1 C$, $K_2 = G_2 C$, and $K_3 = G_3 C$. Then we have the sphere, S, partitioned into the four disjoint sets of points K_1, K_2, K_3, and P. We can write $S = K_1 \cup K_2 \cup K_3 \cup P$.

Figure 5.6 illustrates how we have the sphere partitioned. Remember, at this point we are partitioning the sphere S, and not the ball B. We are partitioning the surface of the ball.

By definition of K_1 and K_2, we see that if all points of K_1 are rotated by the basic rotation τ, we get K_2. So K_1 and K_2 are congruent and we write $K_1 \cong K_2$. We also note that K_3 is obtained from K_1 by rotating K_1 by τ^2. So $K_1 \cong K_3$. And finally, rotating K_1 by σ gives us $K_2 \cup K_3$.

$$S - P = K_1 \cup K_2 \cup K_3$$

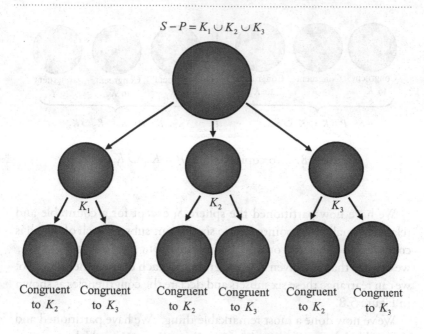

Figure 5.7. Six subsets of $S - P = K_1 \cup K_2 \cup K_3$.

Summarizing: $K_1 \cong K_2 \cong K_3 \cong K_2 \cup K_3$

This last statement is the Hausdorff Paradox. It may not at first appear paradoxical, but consider that the sphere S is made up, almost entirely, of points in $K_1 \cup K_2 \cup K_3$. Since $K_1 \cong K_2 \cong K_3$, we can conclude that each K is roughly *one third* of the entire sphere (or at least one third of $K_1 \cup K_2 \cup K_3$). But if each K is congruent to $K_2 \cup K_3$, we could also conclude that any given K, say K_1, is *one half* of the entire sphere. So which is it, one third or one half? This *half-third* dilemma is the Hausdorff Paradox.

As we shall soon see, the Hausdorff Paradox allows us to separate the sphere S into a finite number of pieces which can be reassembled to form two spheres, each identical in shape to the original.

We will split $S - P = K_1 \cup K_2 \cup K_3$ into two copies of itself and deal with the problem of the poles afterward (Figure 5.7). The technique suggested by Robert French uses $K_2 \cup K_3$ as a *cutting template* which can be placed directly onto each K, partitioning each into two sets, one of which being congruent to K_2 and the other being congruent to K_3. This template will fit perfectly as $K_2 \cup K_3$ is congruent to each K.

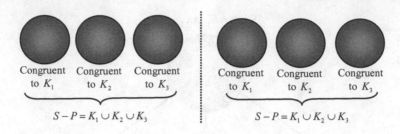

Figure 5.8. Two copies of $S - P = K_1 \cup K_2 \cup K_3$.

We have now partitioned the sphere S, except for a countable and relatively small set of points P, into six disjoint subsets, each of which is congruent to either K_2 or K_3, as illustrated. But since $K_1 \cong K_2 \cong K_3$, we can say that any given K is congruent to each of the other two. So, we can rearrange these six subsets and declare the congruencies as shown in Figure 5.8.

We've now done a most remarkable thing. We have partitioned and reassembled the set of points $S - P = K_1 \cup K_2 \cup K_3$, which represents almost all points on the unit sphere, in such a manner as to form two identical copies of itself. In that $K_1 \cup K_2 \cup K_3$ represents virtually the entire unit sphere S (except for the set of poles P), we are almost done. The fly in the ointment is P.

As it turns out, the problem of the poles is easily disposed of. We can use the set of poles P from the original sphere S to plug the holes representing the missing poles in one of the two copies of $S{-}P$. This gives us one complete copy of S. But how can we fill the holes in the other copy of $S{-}P$ to give us the second copy of the complete sphere? Recall the discussion of piecewise congruence (equidecomposability) given in Chapters 3 and 4. In Chapter 4 we showed that the surface of a sphere missing a countable number of points (holes) is equidecomposable to the complete sphere. We apply this technique, that of shifting from infinity, to the point set $S{-}P$ to give us the second complete copy of the sphere S (Figure 5.9).

This completes the second sphere and we now have two complete copies of the sphere S made from S alone. Call these copies S_1 and S_2 (Figure 5.10).

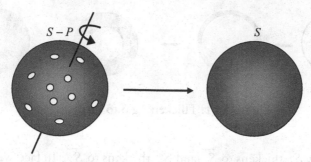

Figure 5.9. Shifting from infinity to show $S - P \sim S$.

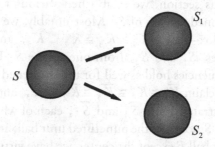

Figure 5.10. Two spheres from one.

Step III—Partitioning the Unit Ball B into Two Copies of Itself

The difficult part of the proof is behind us and it is a rather simple matter to extend the duplication of the sphere S to the duplication of the ball B.

To each subset of S, say K_1 as defined in the previous section, we can associate an inward *thickening*, extending from the surface of the sphere up to, but not including, the center $(0, 0, 0)$. We denote this thickened section of the ball as $\overline{K}_1$. Putting it another way, we can think of K_1 as the projection on S of all points belonging to $\overline{K}_1$. Of course we can thicken all subsets of S in similar fashion.

$$K_1 \text{ thickens to } \overline{K}_1$$
$$K_2 \text{ thickens to } \overline{K}_2$$
$$K_3 \text{ thickens to } \overline{K}_3$$
$$P \text{ thickens to } \overline{P}$$

155

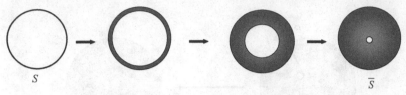

Figure 5.11. Thickening S to form $\overline{S}$.

Similarly, S_1 thickens to $\overline{S_1}$ and S_2 thickens to $\overline{S_2}$. In fact, we could thicken the entire sphere S to $\overline{S}$, the *punctured ball*, as illustrated in Figure 5.11.

In the previous section, we established various relations among the above mentioned subsets of S. Most notably, we established the Hausdorff Paradox—$K_1 \cong K_2 \cong K_3 \cong K_2 \cup K_3$, and the creation of the two spheres, S_1 and S_2, from one sphere S. But, intuitively, all of these congruencies hold as well for the thickened subsets. So we can just as easily claim $\overline{K_1} \cong \overline{K_2} \cong \overline{K_3} \cong \overline{K_2 \cup K_3}$ and that $\overline{S}$ can be separated and rearranged into $\overline{S_1}$ and $\overline{S_2}$, each of which is piecewise congruent to $\overline{S}$. Since $\overline{S}$ is the punctured unit ball, and represents all points of the unit ball B except the center, we have virtually proven the Banach-Tarski Theorem, except for the matter of dealing with the center $(0, 0, 0)$. If we can plug the center holes in each of the two punctured balls $\overline{S_1}$ and $\overline{S_2}$, then we will have created two solid balls from one and our proof will be complete.

We use the center $(0, 0, 0)$ of the original ball B to plug the center hole in $\overline{S_1}$ giving us one complete copy of the ball. We plug the hole in the second punctured ball $\overline{S_2}$ by the now familiar technique of shifting from infinity. That is, we consider the center of the punctured ball $\overline{S_2}$ as one of countably many carefully chosen points on a circle completely contained in $\overline{S_2}$. Shifting the points from infinity plugs the center hole and we now have our second complete copy of B. Call the two copies B_1 and B_2 (Figure 5.12). This completes our proof of the Banach-Tarski Theorem.

We have just proven the *duplication* version of the Banach-Tarski Theorem. The *magnification* or *strong version* of the theorem may be even more striking.

> *If A and B are any two bounded three-dimensional sets with nonempty interiors, then A ~ B.*

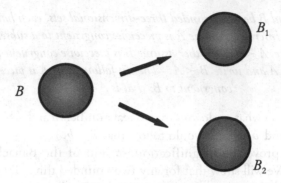

Figure 5.12. Two balls from one.

If, say, A is a ball the size of a pea and B is a ball the size of the sun, then this version of the Banach-Tarski Theorem asserts that the pea (A) can be decomposed into a finite number of pieces and reassembled to form a ball the size of the sun (B) (Figure 5.13).

Interestingly, this version of the theorem does not require that A and B be spherical in shape. In fact, their shapes need not be identical. So, we could just as easily call this version of the theorem the *mosquito and the elephant* paradox!

The magnification version of the theorem follows directly from the duplication version. Before giving the proof, we must state a preliminary theorem—the *Banach-Schröder-Bernstein Theorem*. Its proof is given in this chapter's appendix.

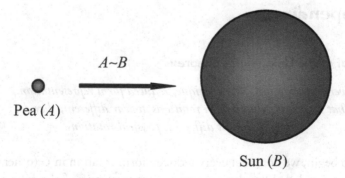

Figure 5.13. The pea and the sun paradox.

Let A and B be two bounded three-dimensional sets, each having nonempty interiors. Assume A is piecewise congruent to a subset of B. We write $A \preceq B$. Further assume B is piecewise congruent to a subset of A and write $B \preceq A$. Then it follows that A is piecewise congruent to B; that is $A \sim B$.

(This is somewhat analogous to two real numbers a and b being such that $a \leq b$ and $b \leq a$. It would follow that $a = b$.)

Now we prove the magnification version of the Banach-Tarski Theorem. We will show that for any two bounded three-dimensional sets A and B having nonempty interiors, $A \preceq B$. Since we could use a similar argument to show $B \preceq A$, it will follow that $A \sim B$.

Choose solid balls V and W such that A is contained in V and W is contained in B. Duplicate W (by Banach-Tarski duplication) until there are enough copies of W that can be translated and overlapped so as to completely cover V. Assume n copies are required. Note that n overlapping copies of W are equidecomposable to a subset of n disjoint copies of W. Furthermore, the Banach-Tarski Theorem guarantees that the n disjoint copies of W are equidecomposable to W.

Consequently,

$A \subseteq V \subseteq n$ overlapping copies of $W \preceq n$ disjoint copies of $W \sim W \subseteq B$,

which establishes $A \preceq B$. Since, by similar means, we can show $B \preceq A$, it follows by the Banach-Schröder-Bernstein Theorem that $A \sim B$. This concludes the proof of the magnification version of the Banach-Tarski Theorem.

Appendix

Proof of the Uniqueness Theorem

Every rotation in G has a unique, reduced form representation. That is, if two reduced form rotations appear different, they do, in fact, represent different physical rotations.

To begin, we note that every reduced form rotation in G (other than $I, \sigma, \tau,$ and τ^2) can be expressed in at least one of the four forms

$$\alpha = \tau^{P_1} \sigma \tau^{P_2} \sigma \cdots \tau^{P_n} \sigma$$

$$\beta = \sigma \tau^{P_1} \sigma \tau^{P_2} \cdots \sigma \tau^{P_n}$$

$$\gamma = \tau^{P_1} \sigma \tau^{P_2} \sigma \cdots \sigma \tau^{P_n}$$

$$\delta = \sigma \tau^{P_1} \sigma \tau^{P_2} \cdots \sigma \tau^{P_n} \sigma$$

where $n \geq 1$ and each exponent P_i is one or two. (For γ, we must have $n > 1$.)

We begin by giving the argument [Osofsky and Adams 78, p. 504] that no reduced form rotation of the form α can equal the identity matrix I. From this we will show (as given by [Stromberg 79, p. 154]) that no reduced form rotation of the form $\beta, \gamma,$ or δ can equal I. After having established that no reduced form rotation (other than I itself) can equal I, we will show that the reduced form representation of any rotation in G is unique.

$$\text{Let } \tau = \begin{bmatrix} -\dfrac{1}{2} & \dfrac{-\sqrt{3}}{2} & 0 \\[2mm] \dfrac{\sqrt{3}}{2} & -\dfrac{1}{2} & 0 \\[2mm] 0 & 0 & 1 \end{bmatrix} \text{ and } \sigma = \begin{bmatrix} 0 & 0 & 1 \\ 0 & -1 & 0 \\ 1 & 0 & 0 \end{bmatrix}$$

be the matrix representations of τ and σ as given in Chapter 3. Then

$$\tau^P \sigma = \frac{1}{2} \begin{bmatrix} 0 & \pm\sqrt{3} & -1 \\ 0 & 1 & \pm\sqrt{3} \\ 2 & 0 & 0 \end{bmatrix}$$

where we take $+\sqrt{3}$ if $P = 1$ and $-\sqrt{3}$ if $P = 2$. It follows by mathematical induction (or empirical verification) that

$$\alpha = \frac{1}{2^n} \begin{bmatrix} m_{1\,1} & m_{1\,2}\sqrt{3} & m_{1\,3} \\[2mm] m_{2\,1}\sqrt{3} & m_{2\,2} & m_{2\,3}\sqrt{3} \\[2mm] m_{3\,1} & m_{3\,2}\sqrt{3} & m_{3\,3} \end{bmatrix}$$

where m_{11}, m_{21}, m_{31}, m_{32}, and m_{33} are even integers and m_{12}, m_{22}, m_{13}, and m_{23} are odd integers. By considering the individual elements of this matrix, it is easy to show $\alpha \neq I$. For example, the fact that m_{12} is odd suggests that

$$\frac{m_{12}\sqrt{3}}{2^n} \neq 0 \text{ and therefore } \alpha \neq I.$$

Now it can be shown that no β rotation can equal I because if it were so, then $\sigma\beta\sigma = \sigma I\sigma = \sigma^2 = I$ which is contradictory because $\sigma\beta\sigma$ is of the form α which we have shown can not equal I.

So now it's on to showing that no γ rotation can equal I. Assume γ is as above with the smallest possible n $(n > 1)$. If $P_1 = P_n$ then either $P_1 = P_n = 1$ or $P_1 = P_n = 2$. In that these two cases are treated similarly, let's assume $P_1 = P_n = 1$. Then $\gamma = \tau\sigma\tau^{P_2}\sigma \cdots \sigma\tau$ and note that $\tau^2\gamma\tau = \sigma\tau^{P_2}\sigma \cdots \sigma\tau^2$, which is of the form β. But if $\gamma = I$ then $\tau^2\gamma\tau = \tau^2 I\tau = \tau^3 = I$ which is certainly not of the form β. So we have a contradiction and $\gamma \neq I$ if $P_1 = P_n = 1$. A similar argument shows $\gamma \neq I$ if $P_1 = P_n = 2$.

To complete the argument that no γ can equal I we consider γ where $P_1 \neq P_n$. Assume $\gamma = I$ and we will arrive at a contradiction. If $n > 3$ then $I = \sigma\tau^{P_n}\gamma\tau^{P_1}\sigma = \tau^{P_2}\sigma \ldots \sigma\tau^{P_{n-1}}$ which is of the form γ but contradicts the choice of the *smallest possible n*. So it must be that $n = 2$ or $n = 3$. If $n = 2$, we have $I = \tau^{P_2}\gamma\tau^{P_1} = \sigma$ which is impossible. And if $n = 3$ we have $I = \sigma\tau^{P_3}\gamma\tau^{P_1}\sigma = \tau^{P_2}$, also impossible. We conclude $\gamma \neq I$.

A δ rotation can not equal I because if this were so then $\sigma\delta\sigma$, which is of the form γ, would equal I and we have shown that that is impossible.

In summary, we have shown that no reduced form rotation in G (other than I itself) can equal the identity rotation I. We continue by establishing that all reduced form representations are unique, representing distinct physical rotations.

Assume the contrary. Let $\lambda_1\lambda_2 \cdots \lambda_m = \rho_1\rho_2 \cdots \rho_n$ represent the equality of two, reduced form rotations (other than I) in G with distinct representations. So, each λ and ρ represents τ, τ^2, or σ. If the representations are distinct then either $m \neq n$ or, for some i, $\lambda_i \neq \rho_i$. Assuming that we can find such an equality of two reduced form rotations, choose two such that $m + n$ is minimal. Recalling from Chapter 3 that the inverse of the rotation $\rho_1\rho_2 \cdots \rho_n$ is given by $(\rho_1\rho_2 \cdots \rho_n)^{-1} = \rho_n^{-1}\rho_{n-1}^{-1} \cdots \rho_1^{-1}$ we claim

$$(\lambda_1\lambda_2 \cdots \lambda_m)(\rho_1\rho_2 \cdots \rho_n)^{-1}$$
$$= \lambda_1\lambda_2 \cdots \lambda_m \rho_n^{-1} \rho_{n-1}^{-1} \cdots \rho_1^{-1}$$
$$= (\rho_1\rho_2 \cdots \rho_n)(\rho_1\rho_2 \cdots \rho_n)^{-1} = I.$$

For these λ and ρ to *collapse* or reduce to I it would be required that $\lambda_m \rho_n^{-1}$ simplify. This is impossible as it would violate the minimality of $m + n$.

Consequently, every rotation in G must have a unique reduced form representation and we have proved the Uniqueness Theorem.

The basic rotations τ (120°) and σ (180°) were defined as to have the angle of 45° between their axes. Hausdorff showed that if $\cos 2\theta$ were transcendental, then θ would suffice as an angle between the axes which would allow for uniqueness. As we have seen, $\theta = 45°$ works as well.

Proof of the First Rotation Equation

To prove that $\tau G_1 = G_2$, we equivalently prove that r is a member of G_1 if and only if τr is a member of G_2. Assume that all elements of G have been sorted into the subsets G_1, G_2, and G_3 as specified by the assignment process given in this chapter. Assume that r is not equal to σ, τ, or τ^2, for if it were so, verification of "r is a member of G_1 if and only if τr is a member of G_2" would be trivial. Note that τr need not have τ as its leftmost character. For example, if the leftmost character of r is τ^2, then the leftmost character of τr is σ.

Case 1. The leftmost character of r is σ.

If r is a member of G_1 then, by the assignment process given in this chapter, τr is a member of G_2. To prove the other direction, assume τr is a member of G_2. Note from the diagram "Partitioning G into G_1, G_2, and G_3" that rotations are assigned to G_2 in only three possible ways. Since τr is a member of G_2 and the leftmost character of r is σ, two of the three possibilities are eliminated and r must belong to G_1. Thus for Case 1, $\tau G_1 = G_2$.

Case 2. The leftmost character of r is τ.

In such a case $r = \tau p$ where p's leftmost character is σ. Assume r is a member of G_1 and note from the diagram that assignments to G_1 are made in only four possible ways. Similar to what occurred for Case 1, we can eliminate three of the four possible assignments and conclude p belongs to G_3. Again using the diagram, $\tau r = \tau^2 p$ is assigned to G_2.

To show the reverse direction, assume $\tau r = \tau^2 p$ is assigned to G_2. As before we note from the assignment diagram that assignments to G_2 are made in only three possible ways. Since p begins with σ, we must conclude p belongs to G_3. The assignment process dictates that $r = \tau p$ be assigned to G_1. Thus, for Case 2, $\tau G_1 = G_2$.

Case 3. The leftmost character of r is τ^2.

In such a case, $r = \tau^2 p$ where p's leftmost character is σ. If r is a member of G_1 and assignments are made to G_1 in only four possible ways, then three of the four possible assignments can be eliminated and we conclude p belongs to G_2. It follows that $\tau r = \tau^3 p = p$ belongs to G_2.

To show the reverse direction, assume $\tau r = \tau^3 p = p$ belongs to G_2. Using the diagram and noting that the leftmost character of p is σ, $r = \tau^2 p$ is assigned to G_1. Thus, for Case 3, $\tau G_1 = G_2$.

The proofs of the second and third rotation equations are similar in form to the proof above and are not presented here.

Proof of the Banach-Schröder-Bernstein Theorem

Let A and B be two bounded three-dimensional sets, each having nonempty interiors. Assume A is piecewise congruent to a subset of B. We write $A \preceq B$. Further assume B is piecewise congruent to a subset of A and write $B \preceq A$. Then it follows that A is piecewise congruent to B $(A \sim B)$.

We duplicate a common proof of set theory's Schröder-Bernstein Theorem. Assume A and B are partitioned as necessary. We are given that the subsets of A (as defined by A's partition), can be made to coincide with some subsets of B (as defined by B's partition) by rigid motions. So there is a one-to-one mapping (injection) of these subsets of A to certain subsets of B. Similarly, there exists a one-to-one mapping (injection) of the subsets of B (as defined by B's partition) to certain subsets of A. Let the function f represent the mapping from A into B (down arrows) and let the function g represent the mapping from B into A (up arrows) (Figure 5.14).

Subsets of A

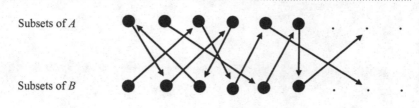

Subsets of B

f is represented by down arrows g is represented by up arrows

Figure 5.14. The Banach-Schröder-Bernstein Theorem.

We wish to show $A \sim B$ and must show that there is a one-to-one and onto correspondence (bijection) between the subsets of A and subsets of B via rigid motions. That is, we must exhibit a one-to-one and onto function (bijection) h mapping subsets of A onto the subsets of B.

Before doing this, we say that a subset (either from A or B) is *B-based* if making backward jumps from the subset terminates in B. If this were not the case, it would either be *A-based* (ending up in A) or would simply not terminate. Note that if a subset is *B-based* then so are all of its ancestors (predecessors) and descendants. Now we proceed to define our function h as follows:

Let a be one of the subsets of A. Define

$$h(a) = \begin{cases} g^{-1}(a) & \text{if } a \text{ is } B\text{-based} \\ f(a) & \text{otherwise.} \end{cases}$$

We show h to be one-to-one and onto. Clearly h is one-to-one since g^{-1} is one-to-one, f is one-to-one, and no *B-based* subset can be mapped to the same subset as a non *B-based* subset. It remains to show that h is onto. Let b be one of the subsets of B. We must show that for any such b there exists a subset a of A such that $h(a) = B$. If b is a *B-based* subset then $g(b)$, a subset of A, is a *B-based* subset and we let $a = g(b)$, for then $h(a)$, $= h[g(b)] = g^{-1}[g(b)] = b$. If b is a non *B-based* subset then $f^{-1}(b)$ is a non *B-based* subset of A and we let $a = f^{-1}(b)$, for then $h(a) = h[f^{-1}(b)] = f[f^{-1}(b)] = b$. Thus, h is both one-to-one and onto; the Banach-Schröder-Bernstein Theorem has been proven.

6 Resolution

One of the more annoying things that mathematicians do is cast doubt upon things that we imagine we understand perfectly well.

—Ian Stewart

When a mathematician or scientist produces a highly counterintuitive result, a resolution is called for. Three options are available, each of which is discussed as it relates to the Banach-Tarski Theorem.

Declare the Result Fallacious

Anyone arriving at unexpected and unbelievable results would suspect an error. The scientist might suspect a design error, statistical error, or an experimental anomaly not typical of the population being investigated. The mathematician suspects an error in computation or logic. If, in fact, there is an error and the error is found, then the entire matter is dismissed as a Type 2 paradox.

Is there an error in the proof of the Banach-Tarski Theorem as presented in the previous chapter? The computations are correct and the logic is flawless. If there is an error, we must look for it in the most vulnerable part of the proof—the application of the Axiom of Choice. Is it correct to employ this controversial axiom if it leads to such a bizarre conclusion?

Critics of the Axiom of Choice point to the conclusion of the Banach-Tarski Theorem as ample evidence as to why the axiom should not be used. They point out that there is no proof of the axiom and argue that we can not declare a choice exists if we are unable to explicitly describe the choice. To simply declare its existence, they say, does not establish, mathematically, its existence.

Defenders of the axiom argue that construction is not required to prove existence. There are many, so called *existence theorems* in mathematics which establish existence of a mathematical object, without actually constructing the object whose existence is guaranteed by the theorem. Axioms, by definition, require no proof. The axiom is highly intuitive and to reject it would be, to some individuals, more disturbing than the consequences of its use. And, in any event, we have been able to produce *good mathematics* with the help of this axiom. It is useful, if not indispensable, in such sub-disciplines as topology, algebra, functional analysis, and real analysis. Why reject it?

The general consensus of the mathematics community is to accept the Axiom of Choice and not send the Banach-Tarski Theorem to the trash bin.

Accept the Theorem at Face Value

If the theorem is not to be rejected, then we are faced with the possibility of accepting it at face value, without further interpretation. This would be a tough pill to swallow as the theorem suggests volumes and masses could be doubled, quadrupled, and in fact magnified to any specified size and shape by a mechanical process of dissection and reassembly.

Historically, it would not be the first time initially unbelievable results are ultimately accepted. As Mark Twain puts it [Twain 97, p. 156], "Truth is stranger than fiction, but it is because fiction is obliged to stick to possibilities; truth isn't." Nicolas Copernicus (1473–1543), the founder of modern astronomy, proposed in 1530 that the earth rotated about its axis once daily and revolved about the sun once yearly—a heliocentric theory directly contradicting the geocentric interpretation of the universe as given by Claudius Ptolemy. The Copernican theory went against the political, philosophical, and religious beliefs of the time, which held that man was the center of the universe and all revolved about the earth. Galileo, a defender of the Copernican theory, was ordered by the powers that be to renounce it and was ultimately condemned to spend the rest of his life in prison. Other defenders, including Johannes Kepler and Sir Isaac Newton, would eventually convince the scientific community of the theory's validity.

Albert Einstein's special theory of relativity makes claims which, at first appearance, are very difficult to accept. One of Einstein's basic postulates is that the speed of light is constant for all observers, regardless of their relative speeds. This is every bit as paradoxical, at least physically, as the Banach-Tarski Theorem. Yet, this is well accepted today.

From the basic postulates of relativity come the Fitzgerald-Lorentz equations

$$L' = L\sqrt{1 - \frac{v^2}{c^2}}$$

$$T' = \frac{T}{\sqrt{1 - \frac{v^2}{c^2}}}$$

$$M' = \frac{M}{\sqrt{1 - \frac{v^2}{c^2}}}$$

In these equations, c denotes the speed of light (constant) and v denotes the speed of an object relative to an observer. Bizarre physical phenomena are mathematically predicted to occur as v approaches c. The first equation tells us that an object of length L will appear shrunk to a length of L', when the object's length is measured by the observer. The degree of contraction depends on the object's speed v. A process taking time T on the object will take a longer time, T', when measured by the observer. The time dilation factor depends, as before, on the object's speed. Finally, the object's mass when measured by the observer will increase to an amount M', again determined by the speed.

As an example, consider an object of length 50 meters and mass 50 kilograms moving at 90% the speed of light (.90c) relative to the observer. When measured by the observer, its length will have shrunk to approximately 22 meters and its mass will have increased to approximately 115 kilograms. An event lasting 20 seconds on the object will appear to take approximately 46 seconds when measured by the observer. Though highly counterintuitive, the validity of the Fitzgerald-Lorentz equations has been verified in laboratory settings.

We could continue with additional examples from relativity theory, quantum mechanics, and even mathematics itself; but, the point is that what appears unbelievable may turn out to be true.

So, do we accept the Banach-Tarski Theorem at face value and proceed to design duplicating machines to increase the world's food supply? Well, not exactly. We hedge a bit here and choose to adopt the third option, as explained below.

Reinterpret the Result

Most mathematicians resolve the Banach-Tarski Theorem as a Type 1 paradox. The conclusion is assumed valid, assuming, as most mathematicians do, that we may invoke the Axiom of Choice. However, the theorem does not allow us to duplicate gold bars and loaves of bread to increase our wealth and feed the world.

To clarify, recall the mathematical concept of *measure*, as given in Chapter 4. The *Lebesgue measure*, or simply *measure*, of a point set quantifies the *extent* of the set. So, we would like the measure of a closed interval of real numbers, $[a, b]$, to be the length of the interval, $b - a$, and the measure of the set of points forming a disc of radius r to be the area of the disc πr^2. Similarly, the measure of a set of points forming a familiar solid should be the volume of that solid. The measure of a single point is zero.

We would like this *measure* to be *finitely additive*. This means the measure of the union of a finite number of disjoint sets is the sum of the measures of the individual sets. We could be more precise about this here, but the reader probably understands the concept. We can think of measure as *size*, or *extent*.

As we have seen in Chapter 4, there are sets of points that are nonmeasurable. These are sets that simply do not have a Lebesgue measure. Despite their existence, there is no measure of their size. And, such was the case in proving the Banach-Tarski Theorem in Chapter 5. The set of points C, the choice set created with the Axiom of Choice, is nonmeasurable. As a result, the sets K_1, K_2, K_3, $\overline{K}_1$, $\overline{K}_2$, and $\overline{K}_3$ are all non-measurable. These K have no definable surface area and the $\overline{K}$ have no definable volume. Their measures are not finitely additive because, to put it bluntly, they have no measures!

What makes the Banach-Tarski Theorem paradoxical is the increase in volume by decomposition and reassembly. It may appear that volumes can be combined in ways to create something from nothing. But, in fact, volumes are not being combined because the subsets associated with the Banach-Tarski Theorem *have no volumes!*

Did we not create two balls from one? In fact, we did! But some of the air is taken out of the paradox once we realize that we are in no way claiming something as absurd as $1 = 1 + 1$. We are simply claiming that the ball can be partitioned and reassembled in a very remarkable way, as a consequence of the Axiom of Choice and the existence of nonmeasurable sets. The Banach-Tarski Theorem confirms the existence of nonmeasurable sets since if the K and $\overline{K}$ were measurable, then the intractable problem of creating something from nothing would arise. Though nonconstructable, Banach-Tarski duplication and magnification is mathematically sound. This is one example of many where mathematics and physics may take their separate paths.

The distinguished theoretical physicist Richard Feynman does not mince words in his book *The Character of Physical Law* when he addresses this issue [Feynman 65, pp. 55–56].

> I should like to say a few things on the relation of mathematics and physics which are a little more general. Mathematicians are only dealing with the structure of reasoning, and they do not really care what they are talking about. They do not even need to *know* what they are talking about, or, as they themselves say, whether what they say is true. I will explain that. You state the axioms, such-and-such is so, and such-and-such is so. What then? The logic can be carried out without knowing what the such-and-such words mean. If the statements about the axioms are carefully formulated and complete enough, it is not necessary for the man who is doing the reasoning to have any knowledge of the meaning of the words in order to deduce new conclusions in the same language. If I use the word triangle in one of the axioms there will be a statement about triangles in the conclusion, whereas the man who is doing the reasoning may not know what a triangle is. But I can read his reasoning back and say "Triangle, that is just a three-sided what-have-you, which is so and so," and then I know his new facts. In other words, mathematicians prepare abstract reasoning ready to be used if you have a set of axioms about the real world. But the physicist has meaning to all his phrases. That is a very important thing that a lot of people who come to physics by way of mathematics do not appreciate. Physics is not mathematics and mathematics is not physics. One helps the other. But in physics you have to have an understanding of the connection of words with

the real world. It is necessary at the end to translate what you have figured out into English, into the world, into the blocks of copper and glass that you are going to do the experiment with. Only in that way can you find out whether the consequences are true. This is a problem which is not a problem of mathematics at all.

The Axiom of Choice is an axiom of mathematics, not of physics. Despite its physical appeal, it makes no physical guarantees. The Hausdorff Paradox and the Banach-Tarski Theorem, which rely on this axiom, are true mathematically, in the sense that the reasoning which leads to these results is consistent. The pattern of *assumption→ deduction → conclusion* is being followed. But conclusions so arrived at need not be true in any physical sense.

A diagram similar to the one in Figure 6.1 appears in *The Mathematical Experience*, by Philip J. Davis and Reuben Hersh [Davis and Hersh 81, p. 129].

In the case of Banach-Tarski, we may be unable to make the "Implication to Real World" step shown on the bottom of the chart. This is so because we are unable to physically execute the constructions suggested by the theorem. There are no mathematical inconsistencies here; the process breaks down when we try to extend the mathematics to specific physical situations.

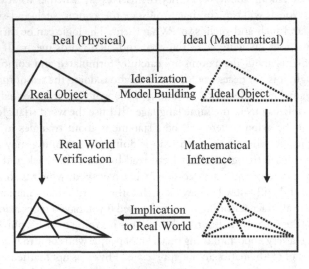

Figure 6.1. Mathematical modeling.
(Reprinted with the permission of Springer-Verlag.)

The idea is well stated by 1967 Nobel Prize Laureate in Chemistry, Manfred Eigen [Mehra 73, p. 618]. He notes, "A theory has only the alternative of being right or wrong. A model has a third possibility; it may be right but irrelevant." And such may be the case when we try to physically duplicate and magnify via the Banach-Tarski Theorem.

Clearly, the Banach-Tarski Theorem is of extreme mathematical significance. But is it physically useful? There is no requirement that it be so. The great mathematician G. H. Hardy (1877–1947) writes of his own mathematical discoveries [Hardy 67, pp. 150–151],

> I have never done anything "useful". No discovery of mine has made, or is likely to make, directly or indirectly, for good or ill, the least difference to the amenity of the world. I have helped to train other mathematicians, but mathematicians of the same kind as myself, and their work has been, so far at any rate as I have helped them to it, as useless as my own.

This self-criticism may be a bit harsh. The point being made is that pure mathematics, by definition, need not extend to the real world. Some pure mathematicians will go so far as to prefer it not be applied, as application of the art in some way soils its beauty.

So, does mathematics really deserve the honor of being called *Queen of the Sciences*? It surely does! Despite some instances where physical application may not exist, mathematics has been the primary tool, historically, of the social, life, and physical sciences. It is remarkable that a study, so potentially pure, can be so applicable to everyday life.

Albert Einstein questions [Einstein 83, p. 28], "How can it be that mathematics, being after all a product of human thought which is independent of experience, is so admirably appropriate to the objects of reality?" This striking duality gives mathematics both power and charm.

Can we rule out, with 100% certainty, the possibility of there ever being a physical application of this theorem? The answer may surprise you.

7 The Real World

I have met people who found it astonishing that cats have holes in their fur at exactly the places where the eyes are.

—Georg Christoph Lichtenberg

Years ago my grandparents had given me a fishing reel. After some use it became dirty and covered with fish scales. I decided to disassemble and clean it. After cleaning, I tried to reassemble what must have been a thousand nuts, bolts, gears, springs, etc. I did it, except for the fact that I had enough parts left over to make another reel.

It was a *Banach-Tarski moment*!

Now I sit at the computer, composing this manuscript and note, with annoyance, dog hair on the keyboard, on my clothes, floating in my coffee, everywhere.

Our Lab, Bailey, sheds so much hair! Another Lab owner suggests we collect the hair from the vacuum cleaner bag and, with it, construct a whole new Lab.

Does this require the Axiom of Choice?

So, of what *real world* use is the Banach-Tarski Theorem? As the theorem is not even a century old, we could adopt the "wait and see" attitude. We are reminded of the cliché, "What use is a newborn baby?"

The nonconstructible nature of the Banach-Tarski decomposition eliminates all possibility of duplicating bars of gold, loaves of bread, etc. But we should not close the door on the possibility of real world application of this surreal result. There may yet exist some wiggle room when it comes to the paradox manifesting itself in real world phenomena. *We* may not be able to actively duplicate and magnify as allowed by the Banach-Tarski Theorem; but let's not impose our shortcomings on *Mother Nature*. She may not have chosen to play by our rules!

As we consider the scenarios, we must distinguish between fantasy, speculation, and reality. The borders are not always clearly marked.

Fantasy

Some years ago I came across the article entitled "A Matter Fabricator Provides Matter for Thought" [Dewdney 89], appearing in the Computer Recreations column of the April, 1989 edition of *Scientific American*. The columnist, A. K. Dewdney, is a long-time contributor to *Scientific American* and a prolific author of books on mathematics and computer science. The article, no more than a few pages in length, may have gone unnoticed by many readers.

After glancing at the title, I assumed the article was to be about *digital fabricators*, also known as *fabbers*, *replicators* and *three-dimensional printers*. Such devices, having been around since the late 1980s, are not to be equated with the fictional replicator of Star Trek fame, a device which could create anything one desired with the press of a button. Digital fabricators are, in fact, real devices that *print* three-dimensional objects from computer data and raw material. In the same sense that a two-dimensional printer puts ink on paper to create a two-dimensional image, the three-dimensional printer sprays layers of plastic powder with a binding agent to form three-dimensional solid objects which may represent models (architectural, automotive, etc.) or actual working devices (Figure 7.1).

Current applications include visualization of scientific data and the creation of teaching aids such as models of the human anatomy. Sony, Adidas, and BMW have found 3D printing technology to be a relatively inexpensive method of producing prototypes.

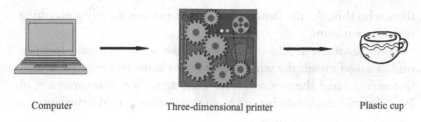

Computer	Three-dimensional printer	Plastic cup

Figure 7.1. Digital fabrication.

Most three-dimensional printers on the market today are industrial, costing well over $100,000. It is expected that less expensive desktop models, for personal use, will appear on the market within the next twenty years. Will there be a market for personal three-dimensional printers? This is anyone's guess; but some envision the day a consumer will go online to purchase the digital description of a product, then have their desktop three-dimensional printer fabricate the object from the digital specifications. Today objects are made from plastic resin; but in the future, the printers may use alloys to produce light bulbs, microchips, and various household appliances. Perhaps retail photocopy shops will offer three-dimensional printing services as well.

I quickly paged through Dewdney's column, assuming it contained nothing new until the words "Banach-Tarski" caught my eye. I decided to read the four page article carefully.

Dewdney writes of recent correspondence he had with a credible mathematician using the pseudonym Arlo Lipof. Lipof claimed to have written a computer program which gave the precise directions for doubling the size of solid spheres via Banach-Tarski decomposition and reassembly! Now I was hooked and read carefully, word for word.

Lipof "mechanized the proof of the Banach-Tarski paradox" and circumvented the nonconstructive nature of the construction by using some sort of random number generator on his personal computer when the Axiom of Choice was required. The account of a seven month long experiment is given where Lipof succeeded in doubling the diameter of a 12 ounce solid ball of gold. The volume and mass of the gold increased by a factor of eight!

Dewdney reports he lost contact with Lipof, but noticed a slow but steady decline in the price of gold in the months prior to the publication of the column. Dewdney writes, "Perhaps that is the ultimate proof for those who thought the Banach-Tarski Paradox was merely a plaything of mathematicians."

I put down the article and tried to make sense of what I had just read. I asked myself the same question that came to my mind when I first encountered the theorem forty years ago. Was this some sort of joke? I could recall only two jokes, of sorts, relating to the theorem—a limerick and an anagram.

There once were two mathematicians
Who split a lead ball with partitions
Though with only five parts
Our two masters of arts
Reassembled it into munitions!

Anagram of Banach-Tarski:

Banach-Tarski   Banach-Tarski

Ah, stupid me! It was the April edition of *Scientific American*.

Anagram of Arlo Lipof:

April Fool!

Speculation

Why bother? Isn't speculation synonymous with fantasy? No! Let's not sell speculation short. As we see in the remainder of this chapter, speculation may be closer to reality than we believe. Bruno Augenstein, formerly of the Rand Corporation, has written extensively on the links between set theory and physics, with specific references to the Banach-Tarski Paradox. He notes many parallels between the two disciplines, some highly speculative and too numerous to list here.

One fascinating comparison, not directly connected with the Banach-Tarski Theorem, is that of the null set $\varnothing$ in set theory and the vacuum in physics. As part of the discussion of the Burali-Forti Paradox given in Chapter 3, we saw how all natural numbers can be constructed from the null set. This being the case, we could go on to argue that arithmetic, algebra, and indeed all mathematics are built up from the singular construct— the null set. In a sense, it all comes from nothing—a vacuum! It's like pulling a rabbit out of a hat (Figure 7.2).

Figure 7.2. Something from nothing.

Augenstein notes [Augenstein 96, pp. 1771–1773] a parallel development of the physical world from the vacuum, reporting that the "vacuum is a very rich physical medium; that it is by a range of interpretations a ferromagnet, a dialectric, a superconductor, a thermodynamic phase, and supports emission, propagation, and absorption of temporal relationships, prior to the introduction of matter, force, or metrical structure, but with a topological structure sufficient to construct a quantum theory; that unification themes in fundamental physics may require for their completion some suitable notions of the vacuum; and that 'matter' and 'empty space' or vacuum are not totally separate kinds of entity." He then cites five specific aspects of the physical vacuum which may "underpin physics."

1. The Casimir Effect

The Casimir effect, predicted to exist by Dutch physicist Hendrick Casimir in 1948, is a weak force which exists between two parallel plates pushed together in a vacuum. Modern physics suggests that a vacuum contains *virtual photons* (or electromagnetic waves) of all possible wavelengths; therefore, there is energy in the vacuum. Pushing the plates together eliminates some photons (or waves) between the plates, thus creating an imbalance resulting from energy between the plates being slightly less than energy outside of the plates. This imbalance creates a small force, known as the Casimir force, or Casimir effect. Quantum field theory predicts this *vacuum energy* to be infinite, suggesting a vacuum to be a source of so called *zero point energy*.

2. The Physical Vacuum as a Source of Particles

The second aspect of the physical vacuum as noted by Augenstein is that of it being a source of particles. The high energy collision of two protons can result in the creation of additional particles in the form of proton–antiproton pairs. Symbolically, $P + P \rightarrow P + P + n(P + \overline{P})$. The number of pairs, n, depends on the collision energy. So, $2n$ additional particles are created out of the beam energy from the vacuum. Later in this chapter we consider a Banach-Tarski explanation of this process, also put forward by Augenstein.

3. An Accelerated Observer's View of the Vacuum

Augenstein cites discussions of *apparent particles* as seen by an observer accelerating through a vacuum, again suggesting a zero point field as a source of mass.

4. The Inflationary Universe

The inflationary universe, as described by the big bang theory, may be an example of the rich structure of the physical vacuum, in which *everything comes from nothing*.

5. Gravitational Effects

Finally, it is speculated that the vacuum, or zero-point field, may be a source of gravity.

There is a permissive *everything goes* principle which may be applied to the physical sciences. Karl Svozil of the Institute for Theoretical Physics at the Vienna University of Technology calls it the *Go-Go Principle*.

> *Every method and object should be permitted as long*
> *as it is not excluded by the rules. That is, anything*
> *that is not forbidden is allowed.*

Svozil notes the principle is undoubtedly progressive and open minded, though may be unreliable in the sense that it can yield false claims. However, it might prove to be "essential for producing novel results, for the discovery of undiscovered land (Hilbert's paradise). . ." [Svozil 95, p. 1546]. Either way, it is the primary inspiration for the following Banach-Tarski speculations.

Cosmology

Cosmologists, physicists, philosophers, theologians, and indeed all of us have wondered, "Where does everything come from?" The truth will ultimately be derived by faith or a correct backward extrapolation in time to the beginning of the universe. In the 1940s the Russian American physicist George Gamow developed the *big bang* theory that the universe began with a hot explosion of matter and energy 10 to 15 billion years ago. (The name *big bang* was first used by British astronomer Fred Hoyle, who promoted the opposing *steady-state theory* of creation. *Big bang* was used as a derogatory description of Gamow's theory.) Though being the generally accepted explanation of the beginning of the universe, cosmologists still puzzle over unanswered questions. What happened before the big bang? Famed Cambridge theoretical physicist Stephen Hawking compares this to asking "What's north of the North Pole?" What actually happened during the big bang itself? What is the destiny of our universe?

Based on ideas of general relativity, Hawking suggests an *initial singularity* occurred at the big bang. Singularities are breakdowns in space and time where the laws of classical physical are not applicable.

M. S. El Naschie, formerly of the Department of Applied Mathematics and Theoretical Physics at the University of Cambridge, has published several papers which speculate on connections between the Banach-Tarski Theorem and the initial (cosmic) space-time singularity associated with the big bang theory of creation. He proposes [El Naschie 95] that a "Banach-Tarski mechanism" is responsible for the vacuum and matter created at the initial singularity. He suggests the Banach-Tarski expansion occurred then and only then because the usual laws of physics and geometry do not exist at singularities, where the concept of volume has no real meaning. He notes ". . . starting from an arbitrary small four-dimensional 'space-time' sphere a Banach-Tarski decomposition would cause a doubling mechanism to set on and give rise to a 'big bang-like' inflation leading to the creation of space-time itself." Similar to Augenstein's ideas on the nature of the vacuum, El Naschie considers the vacuum as an "insubstantial something" in comparison to "nothingness" which he regards as "insubstantial nothing." The vacuum is to be considered as a "rich and complex entity for a starting point of theoretical physics."

Ultimately our universe may experience, according to El Naschie, the reverse process in which the universe collapses to a black hole singularity.

Figure 7.3. El Naschie's cosmology.

Referring to this as the "big crunch," El Naschie envisions the collapse as a reverse Banach-Tarski compression (Figure 7.3).

El Naschie makes it a point to stress he is not suggesting this is what really happened. So then what exactly is his point? He writes that we can not ignore the possibility that this or something similar could have happened. So the Go-Go principle keeps the options open.

Chaos

A large glob of taffy is attached to the two spinning rods of the taffy pulling machine. As the rods spin, the taffy is rhythmically stretched and folded, like the kneading of dough. Flecks of color in the taffy spread out randomly and eventually fill out the entire volume of taffy. It's similar to what we see when we add a teaspoon of white cream to a cup of black coffee. Rhythmic stirring spreads the white cream throughout the coffee in randomly appearing chaos. The apparent random and unpredictable behavior of the color in the taffy and the cream in the coffee are examples of *nonlinear dynamics*, a science popularly referred to as *chaos theory*. Applications exist in meteorology, physiology, hydrodynamics, finance, sociology, and political science. We first discuss the distinction between classical dynamical systems and chaotic dynamical systems; then we present Banach-Tarski applications to chaos theory by M. S. El Naschie and Karl Svozil.

A dynamical system, be it classical or chaotic, can be completely described in terms of the associated *phase space*, a hypothetical space of dimension equal to the number of variables required to specify the state of the system at any given point in time. As an example, consider the motion of a frictionless pendulum, not losing energy as it swings back and forth (Figure 7.4).

At any point in time, the state of the system is completely specified by its position and velocity, a point in phase space. As time progresses, the point traces out an orbit in phase space, representing the chronological

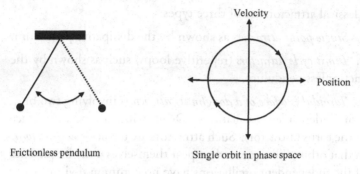

Frictionless pendulum   Single orbit in phase space

Figure 7.4. The motion of a frictionless pendulum.

evolution of the system. One such orbit is given by the circle on the right. If the pendulum described above were dissipative (losing energy due to friction) then the orbit would spiral into a point, as shown in Figure 7.5.

More generally, the state of a particle in three-dimensional space can be specified by six variables—three for the x, y, and z coordinates of its position and three for the x, y, and z components of its velocity. So, a dynamical system involving n particles moving in three-dimensional space would require a $6n$-dimensional phase space.

In the case of the dissipative pendulum, the system eventually comes to rest. For any initial condition the orbit would be attracted to the fixed point in the phase space, corresponding to the origin. Roughly speaking, a subset of phase space to which orbits are attracted is called an *attractor*. No orbits can emanate from an attractor.

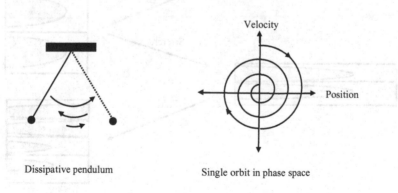

Dissipative pendulum   Single orbit in phase space

Figure 7.5. The motion of a dissipative pendulum.

Classical attractors are of three types:

1. *Single point attractors* as shown by the dissipative pendulum.

2. *Limit cycle attractors* (repetitive loops) such as shown by the frictionless pendulum.

3. *Toroidal (surface of a doughnut) attractors* involving two or more independent oscillations. Orbits wind around the surface of the torus (attractor). Such attractors are termed *quasi-periodic* in that orbits will not exactly repeat themselves if the frequencies of the independent oscillations have no common divisor.

For all classical dynamical systems, orbits which start close together stay close together allowing for long-term predictability of the system. That is, even though we can not measure both position and velocity of a particle with 100% accuracy, a close approximation will allow us to predict a future state with reasonable accuracy.

In 1963, while modeling the earth's atmosphere, meteorologist Edward N. Lorenz of the Massachusetts Institute of Technology discovered another type of attractor—the *chaotic*, or *strange attractor*. In chaotic dynamical systems phase space is mixed like taffy. Repeated

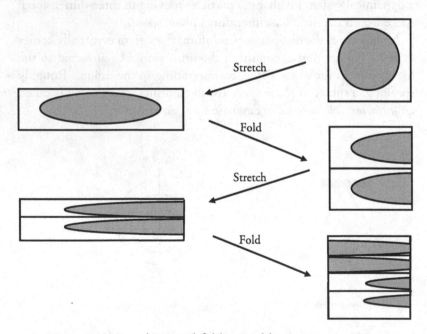

Figure 7.6. Stretching and folding yields a strange attractor.

stretching and folding creates an attractor which is fractal like in the sense that any power of magnification reveals a structure identical to the whole attractor (Figure 7.6). Cross sections of strange attractors are Cantor Sets, as discussed in Chapter 4. Long-term predictions are impossible due to sensitive dependence on initial conditions and the inevitable inaccuracy in measuring the variables associated with the state of the system. Orbits will diverge and fill out the attractor as flecks of color in the taffy or drops of cream in the coffee will eventually fill out their respective spaces. Small initial uncertainties become magnified with time and effectively there is no connection between present and future. Chaos!

In summary, classical dynamical systems allow for predictability because orbits which start close together stay close together. A small error in measurement will at worst create a small error in the prediction of a future state of the system. In contrast, chaotic dynamical systems are highly sensitive to initial conditions; therefore, a small error in measurement may result in a large error with respect to a predicted future state. This magnification associated with chaotic systems is commonly referred to as the *butterfly effect*. According to Lorenz [Lorenz 93, p. 14], the origin of this expression may be Lorenz's paper entitled "Does the Flap of a Butterfly's Wings in Brazil Set Off a Tornado in Texas?"

Both El Naschie and Svozil speculate sources of chaotic behavior in terms of Banach-Tarski processes. El Naschie writes [El Naschie 95b, p. 1507], "The fact that the two spheres which could be produced from a single sphere according to the BT theorem may have arbitrary volumes implies now also the possibility of spatial randomness or chaos invading the universe. If this conjecture is correct then this may indeed be the original source of all randomness and chaos in the world and may be more fundamental than all other 'routes to deterministic chaos' discussed so far in the literature."

Svozil notes ([Svozil 93, p. 222], [Svozil 95, pp. 1550–1553], and [Svozil 96, pp. 789–791]) two comparisons between Banach-Tarski processes and chaotic attractors. In classical, non-dissipative systems, time evolution is distance preserving (an isometry). That is, the distance between two chosen points of phase space remains constant as the points traverse their respective orbits. For example, the distance between any two points chosen from the phase space circle of the frictionless pendulum remains constant as the points move around the circle. Consequently, volumes in phase space remain constant with no apparent possibility of chaotic behavior. Subsets of chaotic attractors spread out in time to fill out the attractor. Svozil proposes an alternate definition of strange attractor as an attractor capable of Banach-Tarski magnification via time evolution. Strange attractors are transformed to fill out the attractor in a manner similar to subsets of a sphere being rearranged to form a sphere of larger volume.

Svozil suggests the possibility of *linear chaos*, claiming there may be routes to chaos from non-chaotic linear dynamical systems. The Banach-Tarski Theorem shows isometric transformations (time evolutions) exist which can double the volume of any subset of phase space. Consequently the attractor of a non-chaotic system can be decomposed and rearranged via carefully chosen time evolutions to an attractor of larger volume. Svozil calls this *linear chaos*. Svozil admits that each transformation required corresponds to a different time evolution; however, he believes there may be a "route to chaos" as a consequence of the volume magnification.

Particle Physics

It used to be simple! My seventh grade science project was an oxygen atom, constructed out of 24 ping pong balls—eight protons painted blue, eight neutrons painted white, and eight electrons painted orange. The protons and neutrons were glued together as the atom's nucleus and the electrons were held by wires so as to orbit about the nucleus. My understanding then was that these ping pong balls represented the fundamental building blocks of matter.

Today's *Standard Model* of particle physics, formulated in the 1970s and experimentally confirmed in the 1980s, contains hundreds of fundamental particles having such characteristics as *mass*, *charge*, *flavor*, and *color*. So, it's no longer the physics of my seventh grade science class!

In today's Standard Model, the primary fundamental particle of matter is the *quark*, which comes in six distinct *flavors*—*up*, *down*, *charm*, *strange*, *top*, and *bottom*—each having its own mass and electric charge. Quarks joined by the so-called *strong force* field form *hadrons*, among which are *neutrons* (one up quark – two down quarks), *protons* (two up quarks – one down quark), and several varieties of *mesons* (one quark – one antiquark). Quarks appear to be truly fundamental and have no substructure. *Leptons*, which include the electron, muon, and neutrino, are also fundamental with no substructure. So we can say that quarks and leptons, including their corresponding antiparticles, are to be considered the fundamental building blocks of matter.

Roger S. Jones, professor emeritus of physics at the University of Minnesota, and Bruno Augenstein use the Banach-Tarski Theorem as an explanation for various quantum phenomena. Jones describes what he refers to as the "electron-muon puzzle" in terms of Banach-Tarski magnification. Electrons and muons, both members of the lepton family of fundamental particles, are alike in all respects except for mass and stability. The muon, *electron-like* and immune to strong interactions, is 200 times as massive as the electron and is not as stable; the lifetime of the muon is about one-millionth of one second. Other than mass and stability characteristics, the particles are identical. The muon may be negatively or positively charged, as is the case with the electron and its positively charged antiparticle—the positron. The negatively charged muon can be rightly called a *heavy electron*.

According to Jones, the muon is unique as an elementary particle in that it has no significant characteristics, other than being a "fat, short-lived electron." Other elementary particles serve a unique purpose, each with a specific function. Jones questions the muon's existence, suggesting that it may have evolved from the electron by Banach-Tarski magnification. He writes [Jones 82, p. 227], ". . . , a pea may be taken apart, like a three-dimensional jigsaw puzzle and put back together to form the sun! And an electron may be transformed into a muon in a finite number of steps . . . ." Jones readily admits to the speculative nature of his discussion, offering the speculation as "food for thought."

The Banach-Tarski Theorem allows for duplication of the solid sphere using a finite decomposition; the theorem itself does not specify the minimum number of pieces required. In 1947, R. M. Robinson [Robinson 47] proved that the Banach-Tarski duplication can be achieved using five pieces, with five being the minimum number of pieces required for duplication. Robinson's decomposition uses two of the five pieces to form one copy and the other three pieces are used to make the second copy. Of course *the process can be repeated by decomposing both the two piece copy and three piece copy into two piece and three piece copies of copies, and so on.*

Augenstein uses the Banach-Tarski Theorem and Robinson's result to explain several hadronic reactions of particle physics [Augenstein 96, pp. 1791–1792]. Hadrons are associated with the solid spheres of the Banach-Tarski Theorem; quarks are the pieces of the decompositions. Three examples follow.

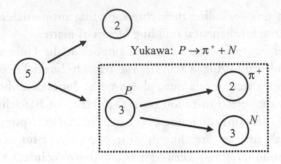

Figure 7.7. BT interpretation of a Yukawa reaction.

Augenstein draws the comparison between the Robinson decomposition and a *Yukawa reaction*, in which a proton (three quarks) changes into a positively charged pi meson (two quarks) and a neutron (three quarks). The Banach-Tarski decomposition of Robinson gives a precise mathematical model of this reaction (Figure 7.7). Is this coincidental, or is Banach-Tarski duplication the source of the *two hadrons from one* Yukawa reaction?

Augenstein's second example, cited earlier in this chapter, is the collision of two protons creating one or more proton-antiproton pairs: $P + P \rightarrow P + P + n(P + \overline{P})$, where the number of pairs, n, depends on the collision energy. Using $n = 1$ the essential part of the reaction is $P \rightarrow P + P + \overline{P}$. When cited earlier, the interpretation given was that the proton-antiproton pair was created from the physical vacuum. Augenstein provides a second interpretation using a result of Jan Mycielski [Mycielski 55] that a solid sphere can be cut into $2 + 3(l - 1)$ pieces and reassembled to form one two piece copy and $l - 1$ three piece copies. Using $l = 4$, Mycielski's result allows the decomposition of a solid sphere into 11 pieces to form one two piece copy and three three piece copies (Figure 7.8). As always, the process can be repeated on any of the newly created spheres.

So repeating the process (using $l = 4$), from one of the three piece copies we can form one two piece copy and three three piece copies. To this we associate the reaction of a proton (three quarks) changing into one neutral pi meson (two quarks), two protons (three quarks each), and one antiproton (three quarks). Symbolically, $P \rightarrow \pi^0 + P + P + \overline{P}$. The symmetric nature of Robinson's result allows for a two piece particle (π^0) to combine with a three piece particle (P), yielding a three piece

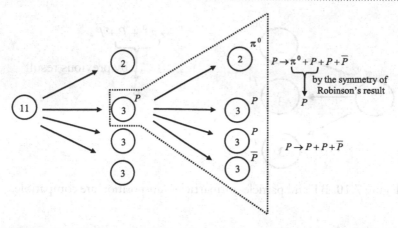

Figure 7.8. BT interpretation of proton-antiproton pair production.

particle (P). We write $\pi^0 + P \rightarrow P$. Putting it all together, we have $P \rightarrow P + P + \overline{P}$, the desired result.

Our last example of the Banach-Tarski Theorem as it may govern hadron behavior involves the transposition behavior of hadronic reactions. Any general reaction is equivalent to one obtained by transposing a particle to the opposite side of the reaction and replacing it with its antiparticle. For example, $A + B \rightarrow C + D$ can be written as $A + B + \overline{C} \rightarrow D$. The Banach-Tarski Theorem along with the results of Robinson and Mycielski are compatible with this. For example, Robinson's result allows (by symmetry) Figure 7.9.

Mycielski's result allows Figure 7.10.

So Banach-Tarski decomposition is compatible with particle-antiparticle transposition as evidenced by the reactions $\pi^0 + P \rightarrow P$ and $\pi^0 \rightarrow P + \overline{P}$.

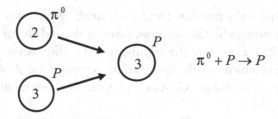

Figure 7.9. Robinson's single sphere from two spheres.

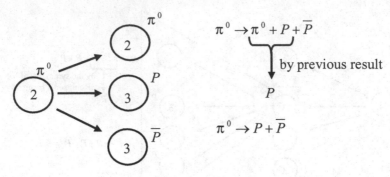

Figure 7.10. BT and particle-antiparticle transposition are compatible.

In summary, Augenstein notes [Augenstein 84, p. 1204], "How one should ultimately regard the significance of these analogies is unclear . . . . But the analogies could still simply reflect fantastic coincidences; or, they could suggest some unsuspected aspects of physical reality." He believes that all coincidences are worth noting as they may yield something significant. He believes the Banach-Tarski Paradox to be both necessary and sufficient to account for hadron behavior. He notes that in either event, such speculations may be tossed out if the connections are, in fact, nothing more than coincidental. Augenstein writes [Augenstein 94, p. 26], ". . . I believe further study of BTT (Banach-Tarski theorems) and reactions to its results may provide fresh insights into areas interesting other audiences. BTT clearly has the power to describe parts of physics. There is reason to speculate further that BTT can shed light on other wide-ranging questions of cognition and learning—how we develop views of physical space, how we fashion and modify beliefs, how relative the notions of 'plausible', 'counterintuitive' are."

Reality

In this chapter we've fantasized and speculated. Which way to reality? At risk of sounding a bit *Clintonesque*, this may depend on our definition of reality. We present here the argument that reality, as we know it, is speculative; in which case the speculations presented are all *real*, in that they do not violate Svozil's Go-Go Principle. We have no justification for rejecting them.

Think about it. Albert Einstein said, "Reality is merely an illusion, albeit a very persistent one." Lily Tomlin, playing the character Trudy in Jane Wagner's Broadway play *The Search for Signs of Intelligent Life in the Universe*, gives her take on reality in saying [Wagner 86, p. 18], "I refuse to be intimidated by reality anymore. After all, what is reality anyway? Nothin' but a collective hunch."

Could it be that any model of the physical world is no more than a way of looking at things—a metaphor? Author and speaker Dr. Wayne Dyer claims

> *When you change the way you look at things,*
> *the things you look at change!*

The following words of John Gribbin, used to describe the potential reality of the various interpretations of quantum mechanics, apply here as well [Gribbin 95, p. 245]. "I stress, again, that *all* such interpretations are myths, crutches to help us imagine what is going on at the quantum level and to make testable predictions. They are not, any of them, uniquely 'the truth'; rather, they are *all* 'real', even where they disagree with one another."

In other words, it may be that there is no *real reality*. Our reality is created from speculation; in fact, our reality is speculation at best! Augenstein and others suggest that physics evolves from a mathematical and cultural context. Augenstein goes so far as to suggest physicists are capable of finding real world counterparts to any mathematical theory and recommends such models be actively sought out. Gribbin refers to this as the *prescient power of mathematics*.

Andrew Pickering, formerly of the University of Edinburgh, promotes a similar philosophy in his book *Constructing Quarks—A Sociological History of Particle Physics*. He suggests [Pickering 84, p. 8] that scientists "make their own history" from their sociological and cultural environment, as opposed to the commonly held view of scientists being "passive mouthpieces of nature," observing, recording and reporting. He acknowledges that experimental results are key to any physical theory but notes such results do not force the scientists hand. Experiments are conducted in a context where judgments are made, and experimental error is to be considered.

Roger Jones, in his book, *Physics as Metaphor* [Jones 82], describes physics as a creative process, going against the commonly held viewpoint that objective measurement determines the nature of the real world. The "cardinal metaphors" of space, time, matter, and number, are seen by Jones as creations of the mind. He favors a more creative role by scientists.

The philosophies of El Naschie, Augenstein, Svozil, Gibbin, Pickering, and Jones suggest Banach-Tarski models of physical reality that are no less conceivable than today's accepted models. But to arrive at such realities, we must proceed, creatively, with the Go-Go philosophy of investigation and trust in the prescient power of mathematics.

Bertrand Russell acknowledges the significance of mathematical power saying [Barrow 88, p. 279], "Physics is mathematical not because we know so much about the physical world, but. . . so little. It is only its mathematical properties we can discover."

Martin Gardner takes it further [Gardner and Fennel 94, p. 9],

> Mathematics is not only real, but it is the only reality. That is that entire universe is made of matter, obviously. And matter is made of particles. It's made of electrons and neutrons and protons. So the entire universe is made out of particles. Now what are the particles made out of? They're not made out of anything. The only thing you can say about the reality of an electron is to cite its mathematical properties. So there's a sense in which matter has completely dissolved and what is left is just a mathematical structure.

Let's be clear. There are no knives sharp enough to perform the dissections required for Banach-Tarski phenomena. Given the fact that infinitesimally minute detail is required, with no means (mathematical or physical) of constructing the pieces, we can safely rule out all hope of performing the duplications and magnifications ourselves. Neither you, I, nor Arlo Lipof will ever succeed in this regard. But Banach-Tarski duplication and magnification may occur naturally, *if we choose to believe so.* To accept such phenomena, we must accept the Axiom of Choice, adopt a permissive Go-Go philosophy with respect to physics, and be willing to define physical reality in such a way that Banach-Tarski processes are being executed. This is not science fiction; it is just one way of making sense of the physical world.

8 Yesterday, Today, and Tomorrow

Prediction is very difficult, especially about the future.

—Niels Bohr (1885-1962)

(Similar thoughts have been expressed by Yogi Berra.)

The story concludes with a beginning. There are no proper endings to mathematical stories—only new beginnings. In this respect mathematics distinguishes itself from the natural and physical sciences, where theories and models of the times come to a proper end when replaced with new theories. In contrast, mathematical theorems rarely become obsolete, as mathematical truth is eternal. Every new theorem or solution signals the beginning of new conjectures, investigations, theorems to prove, and problems to solve. The Banach-Tarski Theorem will remain true and serve as a potential source of mathematics and model of the physical world. Despite the paradox, its truth will endure. As the period of a sentence signals the beginning of a new sentence, it is my hope that the reader's conclusion of this book marks the beginning of the reader's investigation of more mathematical mysteries.

To speculate as to what lies ahead, we look back and briefly summarize the past two centuries of mathematics. Your guess is as good as mine regarding twenty-first century mathematics. I offer my thoughts at the end of this chapter.

Yesterday

The age of most physical and natural sciences can be measured in hundreds of years, but mathematical development requires thousands of years to chronicle, by most accounts beginning with the Egyptians (3000 BC) followed by the Babylonians (1000 BC) and ancient Greece (500 BC). And, with virtually no obsolescence, the body of knowledge grows and evolves affording no hope of any one individual being on the cutting edge of all today's discoveries. Mathematicians carve out their niche and specialize. Today's high school graduate with a relatively strong background in mathematics (algebra, geometry, trigonometry) is still approximately 400 years behind the cutting edge of today's mathematics!

Nineteenth century mathematics included advances in calculus triggering the development of mathematical physics, relativity theory, quantum mechanics, and the nature of matter. Cantor's theory of infinite sets appearing at the end of the nineteenth century extending into the twentieth century created a new paradise for mathematical research.

Most would agree that twentieth century mathematics was launched with David Hilbert's challenge to the mathematical world as his famous list of 23 unsolved problems presented to the International Congress of Mathematicians in Paris. The spirit of the challenge was that there could be no *ignorabimus*. Hilbert's words, "We must know. We shall know." are written on his grave in Göttingen. Gödel's incompleteness theorems of undecidable propositions and incomplete systems closed the book on Hilbert's plea, but at the same time opened up new vistas. *Metamathematics*, a way of looking at arithmetic propositions from the outside rather than within, arose as a new component of mathematical logic. Non-Cantorian set theories would develop without such undecidables as the Axiom of Choice and the Continuum Hypothesis. So the nons and uns of mathematics are not as limiting as they appear.

Twentieth century mathematics saw increased abstraction working its way into analysis, geometry, and algebra. Topology, born of geometry and analysis, grew to dominate mathematical research. Probability and statistics, closely related to set theory, measurable functions, and theories of integration, made advances in response to the sixth challenge on Hilbert's list calling for an axiomatization of the theory of probability.

Today

It is generally accepted that Carl Friedrich Gauss and Henri Poincaré were the last of mathematics' universalists. A century ago it was conceivable for a talented mathematician to have mastered most of the mathematics of the time. Since then, the number of mathematicians has increased several hundred times. Today, with thousands of journals published in over 100 languages, the volume of knowledge is staggering. Estimates of 200,000 new theorems annually, as noted by [Casti 96, p. x] and [Davis and Hersh 81, p. 21] equate to approximately 600 new theorems per day!

On May 24, 2000 at the Collège de France in Paris, the Clay Mathematics Institute of Cambridge, Massachusetts (CMI) laid down the gauntlet for today's mathematicians in presenting the seven "Millennium Prize Problems." It was 100 years earlier that Hilbert gave his famous speech challenging mathematicians of the twentieth century. The CMI's objective was to publicize the great unsolved problems, and not necessarily give a direction to the future of mathematics, as was the case with Hilbert's challenge. In the spirit of celebrating the new millennium, a $7 million prize was designated for what the CMI considered to be the seven most significant mathematical problems which have resisted solution. Rules have been published for the award of $1 million for the solution of any one of the seven problems. Think of the following brief descriptions of the seven problems as a snapshot of today's frontier of mathematical discovery. With each resolution will come new beginnings.

The Riemann Hypothesis

Proposed by Bernhard Riemann in 1859 and listed by Hilbert in 1900 as one of his famous list of 23 problems, this may well be the number one unsolved problem of mathematics today. Riemann observed that the distribution of the prime numbers is related to the *Riemann zeta function*,

$$\zeta(s) = \frac{1}{1^s} + \frac{1}{2^s} + \frac{1}{3^s} + \frac{1}{4^s} + \ldots.$$

The hypothesis is that certain solutions of $\zeta(z) = 0$ also satisfy $\mathrm{Re}(z) = 1/2$; that is, the real part of these solutions is equal to $\frac{1}{2}$. If true, the assertion would shed light on the distribution of the prime numbers

193

with possible applications in physics and modern communications technology.

Yang-Mills Theory and the Mass Gap Hypothesis

Fifty years ago physicists Chen-Nin Yang and Robert Mills developed a new theory describing the elementary particles and the strong interactions of these particles. Quantum Yang-Mills theory is now well accepted experimentally and has been confirmed by computer simulation; however, the mathematical treatment remains incomplete. This millennium challenge calls for the mathematics which would complete the mathematical foundation of the theory. Specifically, a quantum mechanical property called the *mass gap* refers to some quantum particles having positive masses, despite the fact their associated waves travel at the speed of light. The mass gap hypothesis refers to the solution of the so-called *Yang-Mills equations* and helps explain why electrons have mass.

The P Versus NP Problem

The only one of the seven problems about computers, the *P versus NP problem* is about the efficiency of problem solving. Computer scientists characterize computation tasks as being of type(s) P, E, or NP. Type P (polynomial time) computations can be done in a relatively short period of time. Type E (exponential time) might require millions of years. Type NP (non-deterministic polynomial time) problems can be thought of as intermediate—where answers could be quickly checked, but the process of arriving at the answers in a deterministic way might take an astronomically large number of years. Such problems would be effectively impossible to solve.

Are P problems truly distinct from NP problems? Do there exist NP problems which can not be solved in polynomial time, other than by random and luck guesses? The question has remained open for over 30 years. A solution would yield applications in commerce and electronic communications.

The Navier Stokes Equations

The *Navier Stokes equations*, produced in the nineteenth century, describe the turbulent motion of fluids and gases. The challenge is to find a solution to these equations in the form of some general formula, if such a solution exists. Engineers are able to approximate solutions to specific cases but a precise solution would complete the mathematics from which applications in the study of turbulence would certainly follow.

The Poincaré Conjecture

Henri Poincaré raised this question one hundred years ago. An elementary description of the conjecture is given by comparing the surface of an apple to the surface of a donut (torus). Imagine a rubber band, stretched around the surface of an apple. Theoretically the rubber band can be shrunk to a point on the apple's surface by continuously sliding it on the surface until it shrinks to the point. On the other hand, if a similar rubber band were stretched around a donut, it could not be shrunk to a point without cutting the rubber band or the donut. The surface of the apple is *simply connected*, the surface of the donut is not. Poincaré asked if a similar distinction could be extended to surfaces in four-dimensional space.

The Birch and Swinnerton-Dyer Conjecture

There are infinitely many ordered triples of whole numbers (x, y, z) satisfying the equation $x^2 + y^2 = z^2$. Euclid found a formula giving all solutions. In 1994 Andrew Wiles proved Fermat's Last Theorem showing that for $n \geq 3$ there are no whole number solutions to $x^n + y^n = z^n$ other than the trivial solutions involving 0s and 1s. For more complicated cases there is no way to determine when such equations have whole number solutions. In special cases the *Birch and Swinnerton-Dyer Conjecture* gives some information. The conjecture is related to the Riemann Hypothesis; so resolving this problem would contribute to our understanding of the prime numbers and their distribution.

The Hodge Conjecture

The general question associated with the *Hodge Conjecture* asks to what extent complicated shapes can be formed from simpler ones of increasing dimension. Progress on this question helps mathematicians categorize a variety of objects. The Hodge Conjecture asserts that for a particular category called *projective algebraic varieties* the pieces called *Hodge Cycles* are combinations of geometric pieces called *algebraic cycles*.

So there they are, the problems of the new millennium. Solve one and you're a millionaire! Of course there is a review process for submitted solutions including publication of the solution in a recognized mathematics journal of world wide repute and a two-year period of general acceptance. Specific information about the problems and prizes can be found at CMI's website www.claymath.org.

The Banach-Tarski Theorem is not a hot topic of today's mathematical frontier; however, Banach-Tarski type arguments appear in various areas of mathematics and several open issues may require Banach-Tarski arguments for their resolution. In 1991, Randal Dougherty and Matthew Foreman published several results of interest. Their main result was that the Banach-Tarski paradox could be performed using pieces with the *property of Baire* [Dougherty and Foreman 94]. The precise nature of this property is a bit too esoteric to discuss here; however, in a related result they were able to prove a version of the Banach-Tarski paradox which, surprisingly, does not require the Axiom of Choice!

To explain, we need some definitions. A set U is *open* if every point of U is contained strictly within a neighborhood of points, all of which also belong to U. For example, the interior of the unit sphere $\{(x, y, z) : x^2 + y^2 + z^2 < 1\}$ is an open set because every point belonging to this set can be the center of a small neighborhood, or ball, completely contained in the interior of this unit sphere. The same can not be said of the ball $\{(x, y, z) : x^2 + y^2 + z^2 \leq 1\}$ which includes its boundary, because any neighborhood of a boundary point, no matter how small, must extend outside the ball. A set U in a space V is *dense* in V if every neighborhood (ball) in V contains a point of U. For example, the set of points in three dimensional space having rational coordinates is dense in three-dimensional space. The set of points in three-dimensional space with integer coordinates is not dense. In simpler terms, one set is dense in another if it *fills out* the second set.

Dougherty and Foreman proved that for any two nonempty bounded open subsets A and B of n-dimensional space ($n \geq 3$), there is a pair wise disjoint collection of open subsets of A whose union is dense in A, such that they can be reassembled to form an open set dense in B. This is done without the Axiom of Choice. As Foreman describes the result, one could find an open dense subset of a tennis ball, split it into a finite number of open subsets and then reassemble the pieces to form an open dense subset of a stadium. We can't blame the Axiom of Choice for the paradox as it is not used in the construction. With this version, nonconstructability is no longer an issue and the question arises once again regarding possible physical application. Time will tell.

Tomorrow

Which path will the explorer take on a hike through the woods? The decision will depend on *curiosity*, *necessity*, and *opportunity*. Curiosity will direct the hiker to an interesting view or rock formation. The hiker chooses to turn left and hike to the top of hill for a better view of what lies beyond. There may be no practicality to this—just the inquisitiveness of human nature. Other paths may be chosen out of necessity. The hiker may be thirsty and chooses the quick path to the river or tires and chooses the easy trail back to camp. Unexpected bad weather requires a temporary shelter from the storm. An unexpected opportunity may direct the hiker in a third direction. A temporary lull in the storm would give our explorer the chance to advance now while the weather permits.

The curiosity and necessity of our metaphorical hiker can be likened to the pure and applied side of mathematics. A century ago, there was not the distinction between pure and applied mathematics as there is today. Noted mathematician and author Paul Halmos characterizes the difference (as reported by [Stewart 87, p. 224]), describing the pure mathematician as being aware of the distinction, while the applied mathematician believes there is none. The demands of the physical, life, and social sciences will direct the applied mathematician as they do today. Ian Stewart views the distinction between pure and applied as a self-destructive split which he optimistically predicts will be coming to an end. He notes the divisions between the subdisciplines have begun to blur as they become intertwined. Stewart writes [Brockman 02, p. 32]

> Over the next fifty years, the trend toward greater unification will accelerate, and soon there will be just mathematics, with no qualifying adjectives and no sectarian disputes. There will still be specialists, but their specialties will combine the abstract logic and conceptual emphasis of pure mathematics with the concrete concerns of applied mathematics. We will all be mathematicians, all part of the same great endeavor, dabbling in our own little patches of the great collective "extelligence" of mathematical thought.

He adds, "The golden age of mathematics is not ancient Greece or Renaissance Italy or Newtonian England, but now. And in fifty years' time it will still be now."

Once again, we see no endings, just beginnings.

As water flowing downhill takes the path of least resistance, opportunity will plot the course of future research in mathematics. Clearly every new theorem opens doors of opportunity; mathematics is this way by nature. What is unique to our time and immediate future is the computer revolution, which began fifty years ago. New fields of study, including linear programming, game theory, operations research and econometrics, sprung up as a consequence of computational power. The computer gives the mathematician and all scientists opportunities not conceivable centuries prior. Significant calculation and communication power removes obstacles, allowing mathematics to take its course *naturally*, as a direct result of these technological advances. With no end to the revolution in sight, it will be some time before historians can assess the full impact the computer revolution has had on pure and applied mathematics.

Moore's Law, stating that computer power doubles every eighteen months, has been accurate over the past half century. Since the 1950s this power has increased by a factor of ten billion. If we go back eighty years, computer power has increased by a factor of one trillion. The Internet is doubling in size every year, allowing for rapid dissemination, review, and publication of mathematical discoveries via email, newsgroups, and electronic publication. Nobel prizes have been awarded for work whose main methodology is computational. The computer is now an indispensable tool for all the mathematical sciences.

As the world grows smaller through electronic communication, China will be producing mathematicians at a rate proportional to its size. More mathematicians = more mathematics. More beginnings!

To the reader who may believe the revolution will soon run its course, think again. The computer revolution will continue to fuel itself for decades to come. The improving power-to-cost ratio for computer hardware will avail the technology to more private businesses, schools, and individuals. New generations of computer savvy students will be entering universities worldwide, and this will certainly affect the teaching, learning, and doing of mathematics. The computer is already a common household appliance, with microprocessors becoming cheap and plentiful. The computer will be part of our immediate environment, with microprocessors being scattered about as paper is today. The average person will use a computer as casually as we pick up a telephone or turn on the television.

In years to come, expect silicon chip technology to become obsolete. The problem with silicon chips is that the wires within the chips are becoming thinner and electrons can *leak* or *tunnel* across the wires. So there are limits and we should expect new and better technologies. *Optical computers* will use light beams, instead of wires, to carry digital information at the speed of light, generating less heat than conventional microchips. Optical transistors have been developed by scientists at Bell Labs which regulate the flow of light. Don't expect your next PC to be optical, but the optical PC is a legitimate candidate to replace silicon technology in years ahead.

New on the scene are *DNA computers*, some of which have already been built, which can solve some mathematical problems faster than today's supercomputers. Think of a DNA computer as a test tube containing 10^{20} molecules, all calculating simultaneously. DNA computers can calculate on large numbers of molecules simultaneously and are a billion times more energy efficient than silicon chip computers. Consequently, one ounce of DNA could be 100,000 times faster than today's fastest computer. Information is stored digitally, replacing the binary zeros and ones of silicon technology with the four symbols, A, T, C, and G corresponding to the four nucleic acid components of DNA. Storage capability is phenomenal with one cubic centimeter of DNA being capable of storing 10^{21} bits of information, the equivalent of one trillion CDs! A pound of DNA in 1000 quarts of liquid, filling up a cubic yard, would be able to store more memory than all the computers manufactured to date.

As with any technology, there is a downside. DNA decays and can not store for long periods of time. Long term storage would require data be transferred to conventional memory.

The ultimate computer—the *quantum computer*—is now in the planning stages. Wires and circuitry of conventional computers are replaced by quantum waves and the fundamental unit of information is the quantum bit or *qubit*. Such computers would be astronomically more powerful than today's supercomputers. The potential speed of quantum computer computation is almost incomprehensible. Consider that it took 1600 computers connected via the Internet eight months to factor a 129 digit number. Peter Shor of AT&T Bell Laboratories has devised a quantum computer algorithm which could factor a number on the order of 10^{200} digits (yes, digits) in seconds! Applications are in the field of encryption (financial privacy, natural security, etc.) which rely on the inherent difficulty in factoring large composite numbers.

In theory, a quantum computer is capable of perfectly modeling any physical process and applications are limitless.

Again, there is the expected downside. Quantum transistors and computers are highly vulnerable to contamination. A single cosmic ray or atomic particle could corrupt a computation, requiring that such computers be housed and isolated in so called *clean rooms*. Progress in quantum computer technology is slow and it may take a century before such computers become a reality. Today thoughts of quantum computing are a fantasy—the beginning of tomorrow's reality.

But how does all of this affect the theorem-proof nature of pure mathematics, which gave us the Banach-Tarski Theorem? Computer technology, a product of computer science and to some extent pure mathematics, will now repay the favor by serving as a powerful tool for the pure mathematician, not just as a means of computation and communication, but in ways which may change the very way mathematicians mathematize.

A century ago the typical proof of a mathematical theorem was relatively short and authored by a single individual. As the body of mathematical knowledge grows, proofs are becoming lengthier, often written by teams of mathematicians. Andrew Wiles' famous proof of Fermat's Last Theorem is about two hundred pages in length and was originally accepted largely on Wiles' reputation. It is estimated that less than one out of 1000 professional mathematicians are qualified to evaluate Wiles' proof. Though Wiles did not use a computer in his proof, one wonders if there is a limit to the complexity that can be achieved without a computer.

In *The Death of Proof*, John Horgan asks [Horgan 93, p. 93] if "the proof of Fermat's Last Theorem was the last gasp of a dying culture." There may be a limit as to what can be written, reviewed, and tested without computer assistance.

The record holder, so to speak, may be the published classification of the finite simple groups. Published in the 1980s, the complete work is 15,000 pages in length with over 100 contributors. Has any one mathematician read it all word for word? I don't know.

There have been several well known instances of proof relying on computer assistance, most notable of which is the solution to the *Four-Color-Map Problem*, published in 1976 by Kenneth Appel and Wolfgang Haken. The conjecture was that four colors are sufficient to color any planar map so that no two neighboring regions are of the same color

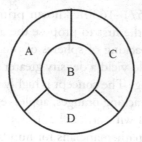

Figure 8.1. A planar map requiring four colors.

(Figure 8.1). Appel and Haken reduced the problem to 2000 special cases requiring 1000 hours of computer time to establish the proof. One must trust the software, hardware, and computer code to accept the proof as it is too lengthy to check manually.

In 1998, Thomas Hales of the University of Michigan solved *Kepler's Sphere Packing Conjecture* by proving that the most efficient (dense) ways of stacking spheres of equal size is by staggering the rows of the first level and then placing the spheres of upper levels in the hollows formed by the level beneath. Depending on how the additional layers are staggered, there are many procedures which produce the optimal density of

$$\frac{\pi}{\sqrt{18}} \approx 74\%.$$

The best known are *cubic close packing (face-centered packing)* and *hexagonal close packing* (Figure 8.2).

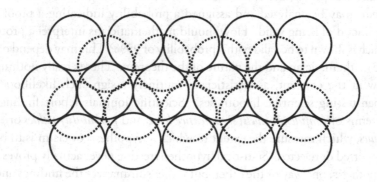

Figure 8.2. Close packing for optimal sphere density—$\dfrac{\pi}{\sqrt{18}} \approx 74\%$.

Johannes Kepler (1571–1630), known primarily for his laws of planetary motion, was the first to propose the above density as being optimal. That is, no packing of spheres of the same radius in three-dimensional space would yield a density greater than that of cubic (or hexagonal) close packing. The concept is highly intuitive and has been used by grocers when stacking oranges, apples, etc. (Hales notes there are problems associated with artichokes!) Yet, a formal proof of the conjecture has eluded mathematicians for hundreds of years. It is the oldest problem of discrete geometry and appears as part of problem number 18 on Hilbert's list.

Published in 1998, Hales' proof consists of 250 pages of text and three gigabytes of computer programs and data. Hales notes [An Update on Hales' Proof . . . 04, p. 3], "nearly every aspect of the proof relies on computer verification." As of this writing, referees have been unable to check the proof by hand; it is simply too extensive. If the details of the proof can not be certified, is it to be accepted? Opinions vary.

Hales is proposing a unique means by which his proof may ultimately be certified. He is attempting to create a *formal proof* of the conjecture in which computer software performs the verification line by line. The effort, known as the *Flyspeck Project* (from *FPK*, an acronym for *Formal Proof of Kepler*) may require up to 20 years to complete. It is highly collaborative and relies on tools (the Internet, computer power) which were nonexistent 50 years ago. Additional information can be obtained at http://www.pitt.edu/~thales/flyspeck/index.html.

Expect more such computer proofs in years ahead. A new twist developed to the computer proof is the *probabilistic proof*, which assigns a statistical probability to the validity of a proof. Lengthy and complex proofs may be analyzed and assigned a probability indicating a proof's likelihood of being valid. How should mathematicians interpret a proof which is shown to be valid with a probability of 99%? This may depend on the mathematician and the nature of the theorem. There's really nothing new in the concept, as statisticians routinely assign such likelihoods when testing scientific hypotheses. Scientific journals report findings in terms of *confidence levels*, *significance levels* and *probability values* or *p-values*, which measure the extent to which a hypothesis or claim is to be supported or rejected. Statistical hypothesis testing never actually proves a hypothesis, one way or the other, but rather summarizes the findings and reports a statistical likelihood of truth. But such ideas are new to pure mathematicians where truth does not come in shades of gray.

In addition to the changing style of proof, there are some mathematicians going so far as to suggest *proof* will lose its status as the gold standard of mathematical truth.

As a direct result of the computer revolution, the traditional and time honored *theorem – proof* way of doing mathematics is undergoing fundamental changes. Computer power, the Internet, and the prevalence of desktop computers have given rise to a new style of doing mathematics known as *experimental mathematics*. Using the computer as a *fast pencil*, the mathematician is able to explore structures, search for patterns and test conjectures in ways previously unimaginable. The experimental mathematician uses the computer to understand mathematical phenomena as the biologist uses the microscope in the laboratory. Visualization of phenomena in cellular automata, chaos theory and applications of fractals would be impossible without the computer. The future mathematician will be on equal footing with other scientists who have benefited from technological assistance for centuries. Departments of experimental mathematics are forming at universities worldwide and the journal *Experimental Mathematics* showcases mathematics inspired by experimentation.

As a tool for visualization, the computer has been used to illustrate the hyperbolic plane interpretation of the Banach-Tarski Theorem, discovered by Stan Wagon and Jan Mycielski in 1984, and presented in the Chapter 1 Appendix of this book. The figures appearing in this book were originally drawn by hand. In Wagon's article "A Hyperbolic Interpretation of the Banach-Tarski Paradox" [Wagon 93a], *Mathematica* software is used to draw the figures and illustrate the group theory algebra behind the paradox. The same software was used by Curtis Bennett to produce his Escher version of the paradox [Bennett 00]. (A brief description of Bennett's article is given in Chapter 1's appendix.) Wagon writes [Wagon 93b, p. xviii], "Of course, computers have not played the same role in this field as in number theory or dynamical systems. Yet they have played some role. . . . In other words, . . . [figures] which were painstakingly discovered and produced by hand can now be created with much less effort by a computer. More exciting, this allows the possibility of carrying on new explorations in this direction."

Experimental mathematics is not without some controversy. Some consider it a passing fad. One traditionalist refers to the above mentioned journal as the *Journal of Unproved Theorems*. The reluctance to change is understandable. Ronald Graham of AT&T Bell Laboratories comments

[Horgan 93, p. 103], "It would be very discouraging if somewhere down the line you could ask a computer if the Riemann Hypothesis is correct and it said, 'Yes, it is true but you won't be able to understand the proof'."

I believe reports of the death of proof are premature. Experimentation leads to intuition and conjecture; it will never replace the certainty associated with proof. To the contrary, conjecture following experimentation will lead to an ever increasing number of new theorems and their proofs. Proof will remain the supreme arbiter of mathematical truth.

The source of mathematics' beauty and power is in its absolute certainty. We should not expect theorems to be replaced by such terminology as *it is likely that* and the *QED* appearing at the conclusion of a proof will not be replaced by *You can bet on it. Trust me!* Proof will never die and there is plenty of room in our world for the experimental, applied, and pure mathematicians to discover, create, and collaborate. It's all just beginning.

By Hilbert's account, an old French mathematician once said, "A mathematical theory is not to be considered complete until you have made it so clear that you can explain it to the first man you meet on the street." By this measure, a long road lies ahead with respect to the Banach-Tarski Theorem and other paradoxes of theoretical mathematics. There's no end in sight, so let's enjoy the journey!

Bibliography

Aczel, A., *The Mystery of the Aleph*, New York: Four Walls Eight Windows, 2000.

Addison, J. W., "Obituary for Alfred Tarski," *California Monthly* **94**(2) (1983), 28.

Albers, D.J., Alexanderson, G.L., and Reid, C., editors, *More Mathematical People: Contemporary Conversations*, Boston: Harcourt Brace Jovanovich, Inc., 1990.

An Update on Hales' Proof of Kepler's Conjecture, *Focus—The Newsletter of the Mathematical Association of America* **24**(5) (2004), 3.

Aquinas, St.T., *Summa Theologiae*, Ia7.4. Quoted in Rucker, R., *Infinity and the Mind*, Boston: Birkhäuser, 1982, 49.

Aristotle, *Physica* III. 207b. Quoted in Boyer, C., *The History of the Calculus and its Conceptual Development*, New York: Dover Publications, Inc., 1949, 41.

Arnold, V., Atiyah, M., Lax, P., and Mazur, B., editors, *Mathematics: Frontiers and Perspectives*, Princeton: International Mathematical Union, 2000.

Augenstein, B., "Hadron Physics and Transfinite Set Theory," *International Journal of Theoretical Physics* **23**(12) (1984), 1197–1205.

———, "Speculative Model of Some Elementary Particle Phenomena," *Speculations in Science and Technology* **17**(1) (1994), 21–26.

———, "Links Between Physics and Set Theory," *Chaos, Solitons and Fractals* **7**(11) (1996), 1761–1798.

Banach, S., and Tarski, A., "Sur la decomposition des ensembles de points en parties respectivement congruents," *Fundamenta Mathematicae* **6** (1924), 244–277.

Barrow J., *The World Within the World*, Oxford: Clarendon Press, 1988.

———, *Pi in the Sky*, Oxford: Clarendon Press, 1992.

————, *The Constants of Nature: From Alpha to Omega—the Numbers That Encode the Deepest Secrets of the Universe*, New York: Pantheon Books, 2002.

Bass, T. and Martin, A., "Road to Ruin," *Discover* **13**(5) (1992), 56–61.

Bell, E. T., *Men of Mathematics*, New York: Simon and Schuster, 1965.

Bennett, C., "A Paradoxical View of Escher's Angels and Devils," *The Mathematical Intelligencer* **22**(3) (2000), 39–46.

Bickel, P. J., Hammel, E. A., and O'Connell, J. W., "Sex Bias in Graduate Admissions: Data from Berkely," *Science* **187**(4175) (1975), 398–404.

Blumenthal, L. M., "A Paradox, A Paradox, A Most Ingenious Paradox," *The American Mathematical Monthly* **47** (1940), 346–353.

Bolzano, B., *Paradoxes of the Infinite* (Translated by Prihonsky, F.), London: Routledge and Kegan Paul Ltd., 1950.

Borel, E., *Leçons Sur La Théorie Des Fonctions*, 1914. Quoted in Moore, G., *Zermelo's Axiom of Choice*, New York: Springer-Verlag, 1982, 188.

Boyer, C., *The History of the Calculus and its Conceptual Development*, New York: Dover Publications, Inc., 1949.

————., *A History of Mathematics*, New York: John Wiley and Sons, Inc., 1968.

Brockman, J., editor, *The Next Fifty Years*, New York: Vintage Books, 2002.

Brown, W., "New-wave Mathematics," *New Scientist* **131**(1780) (1991), 33–37.

Casti, J. L., *Five Golden Rules: Great Theories of 20ᵗʰ Century Mathematics—and Why They Matter*, New York: John Wiley and Sons, Inc., 1996.

Casti, J. L., and DePauli, W., *Gödel*, Cambridge: Perseus Publishing, 2000.

Coffin, S., *The Puzzling World of Polyhedral Dissections*, CD-ROM. Oregonia, OH: Puzzle World Productions, 1998. Available on the World Wide Web (http://www.johnrausch.com/PuzzlingWorld/default.htm).

Corcoran, J., "Review of Alfred Tarski Collected Papers—Volume IV," *Mathematical Reviews*, **91h** (1991), 4127.

Court, N., *Mathematics In Fun and In Earnest*, New York: The Dial Press, 1958.

Croft, H., Falconer, K., and Guy, R., *Unsolved Problems in Geometry*, New York: Springer-Verlag, 1991.

Crutchfield, J., Farmer, J., Packard, N., and Shaw, R., "Chaos," *Scientific American* **255**(6) (1986), 46–57.

Dauben, J., *Georg Cantor—His Mathematics and Philosophy of the Infinite*, Cambridge: Harvard University Press, 1979.

Davis, P. and Hersh, R., *The Mathematical Experience*, Boston: Houghton Mifflin Company, 1981.

Dawson, J., "Gödel and the Limits of Logic," *Scientific American* **280**(6) (1999), 76–81.

Devlin, K., *The Millennium Problems*, New York: Basic Books, 2002.

Dewdney, A. K., "Computer Recreations—A Matter Fabricator Provides Matter for Thought," *Scientific American* **260**(4) (1989), 116–119.

Doty, R., *The Macmillan Encyclopedic Dictionary of Numismatics*, New York: Macmillan Publishing Co., Inc., 1982.

Dougherty, R., and Foreman, M., "Banach-Tarski Paradox Using Pieces with the Property of Baire," *Proceedings of the National Academy of Sciences, USA* **89** (1992), 10726–10728.

———, "Banach-Tarski Decompositions Using Sets with the Property of Baire," *Journal of the American Mathematical Society* 7(1) (1994), 75–124.

Douglis, A., *Ideas in Mathematics*, Philadelphia: W. B. Saunders Company, 1970.

Dubins, L., Hirsch, M. W. and Karush, J., "Scissors Congruence," *Israel Journal of Mathematics* **1** (1963), 239–247.

Einstein, A., "Ether and the Theory of Relativity (An Address delivered on May 5th, 1920 in the University of Leyden)." In *Sidelights on Relativity*, New York: Dover Publications, Inc., 1983.

El Naschie, M. S., "On the Initial Singularity and the Banach-Tarski Theorem," *Chaos, Solitons and Fractals* **5**(7) (1995), 1391–1392.

———, "Banach-Tarski Theorem and Cantorian Micro Space-Time," *Chaos, Solitons and Fractals* **5**(8) (1995), 1503–1508.

Eves, H., *A Survey of Geometry—Volume One*, Boston: Allyn and Bacon, Inc., 1963.

———, *Great Moments in Mathematics After 1650*, Washington DC: The Mathematical Association of America, 1983.

———, *Foundations and Fundamental Concepts of Mathematics*, 3rd ed., Mineola: Dover Publications, Inc., 1990.

Falletta, N., *The Paradoxicon*, New York: John Wiley and Sons, Inc., 1983

Feferman, A., "Alfred Tarski," *American National Biography Online* (http://www.anb.org), February 2000, 1–5.

Feferman, A.B., and Feferman, S., *Alfred Tarski: Life and Logic*, Cambridge, UK: Cambridge University Press, 2004.

Feynman, R., *The Character of Physical Law*, Cambridge, MA: The MIT Press, 1965.

Frederickson, G., *Dissections: Plane & Fancy*, Cambridge: Cambridge University Press, 1997.

French, R., "Multiplier les petite pains avec le Theoreme de Banach-Tarski," *Pour la Science* **112** (1987), 113–118. English version: "The Banach-Tarski Theorem," *The Mathematical Intelligencer* **10**(4) (1988), 21–24.

Galileo, G., *Two New Sciences* (Translated by Drake, S.), Madison: The University of Wisconsin Press, 1974.

Gardner, M., *Mathematics, Magic and Mystery*, New York: Dover Publications, Inc., 1956.

———, *Mathematical Puzzles and Diversions*, New York: Simon and Schuster, 1959.

———, editor, *Mathematical Puzzles of Sam Loyd*, New York: Dover Publications, Inc., 1959.

———, *The Incredible Dr. Matrix*, New York: Charles Scribner's Sons, 1976.

(Gardner, M. and Fennel, B. ?), "Gardner on Gardner: JPBM Communications Award Presentation," *Focus—The Newsletter of the Mathematical Association of America* **14**(6) (1994), 9–33.

Gell-Mann, M., *The Quark and the Jaguar*, New York: W. H. Freeman and Company, 1994.

Givant, S., "A Portrait of Alfred Tarski," *The Mathematical Intelligencer* **13**(3) (1991), 16–32.

Gleick, J., *Chaos—Making of a New Science*, New York: Penguin Books, 1987.

Grattan-Guinness, I., *The Search for Mathematical Roots, 1870–1940*, Princeton: Princeton University Press, 2000.

Gribbin, J., "Pay Attention, Albert Einstein," *New Scientist* **137**(1854) (1993), 28–31.

———, "The Prescient Power of Mathematics," *New Scientist* **141**(1909) (1994), 14.

———, *Schrödinger's Kittens and the Search for Reality*, Boston: Little, Brown and Company, 1995.

Halliwell, J., "Quantum Cosmology and the Creation of the Universe," *Scientific American* **265**(6) (1991), 76–85.

Hausdorff, F., *Foundations of Set Theory*, 1914. Quoted in Kline, M., *The Loss of Certainty*, New York: Oxford University Press, 1980, 204.

Hintikka, J., *On Gödel*, Belmont: Wadsworth Philosopher Series, 2000.

Hofstadter, D., *Gödel, Escher, Bach: an Eternal Golden Braid*, New York: Basic Books, 1979.

Horgan, J., "The Death of Proof," *Scientific American* **269**(4) (1993), 92–103.

Jech, T., *The Axiom of Choice*, Amsterdam: North-Holland, 1973.

Jones, R., *Physics as Metaphor*, Minneapolis: University of Minnesota Press, 1982.

Kaku, M., *Visions: How Science Will Revolutionize the 21ˢᵗ Century*, New York: Doubleday, 1997.

Kałuża, R., *Through a Reporter's Eyes—The Life of Stefan Banach*, Boston: Birkhäuser, 1996.

Kanamori,A., "The Mathematical Development of Set Theory from Cantor to Cohen," *The Bulletin of Symbolic Logic* **2**(1) (1996), 1–71.

Kane, G., "The Dawn of Physics Beyond the Standard Model," *Scientific American* **288**(6) (2003), 68–75.

Klee, V., and Wagon, S., *Old and New Unsolved Problems in Plane Geometry and Number Theory*, Washington, D.C.: The Mathematical Association of America, 1991.

Kline, M., *The Loss of Certainty*, New York: Oxford University Press, 1980.

Laczkovich, M., "Equidecomposability and Discrepancy: A Solution to Tarski's Circle-Squaring Problem," *Journal für die Reine und Angewandte Mathematik* **404** (1990), 77–117.

Lorenz, E., *The Essence of Chaos*, Seattle: University of Washington Press, 1993.

Maor, E., *To Infinity and Beyond*, Princeton: Princeton University Press, 1987.

Mehra, J., editor, *The Physicist's Conception of Nature*, Dordrecht: D. Reidel Publishing Company, 1973.

Moore, G., *Zermelo's Axiom of Choice*, New York: Springer-Verlag, 1982.

———, "Sixty Years After Gödel," *The Mathematical Intelligencer* **13**(3) (1991), 6–11.

Mycielski, J., "On the Paradox of the Sphere," *Fundamenta Mathematicae* **43** (1955), 348–355.

Nagel, E., and Newman, J., *Gödel's Proof*, New York: New York University Press, 2001.

Osofsky, B., and Adams, S., "Problem 6102 and Solution," *The American Mathematical Monthly* **85**(6) (1978), 504.

Penrose, R., *The Emperor's New Mind*, Oxford: Oxford University Press, 1989.

Peterson, I., *Islands of Truth—A Mathematical Mystery Cruise*, New York: W. H. Freeman and Company, 1990.

Pickering, A., *Constructing Quarks—A Sociological History of Particle Physics*, Chicago: The University of Chicago Press, 1984.

Reid, C., *Introduction to Higher Mathematics for the General Reader*, New York: Thomas Y. Crowell Company, 1959.

Richter, C., "Congruence by Dissection of Topological Discs—An Elementary Approach to Tarski's Circle Squaring Problem," *Discrete and Computational Geometry* **28** (2002), 427–442.

Robinson, R. M., "On the Decomposition of Spheres," *Fundamenta Mathematicae*, **34** (1947), 246–260.

Royden, H. L., *Real Analysis*, 2nd ed., Toronto: The Macmillan Company, 1968.

Rucker, R., *Infinity and the Mind*, Boston: Birkhäuser, 1982.

Sangalli, A., "Forum: Who Needs Mathematics?—Arturo Sangalli Searches for a Meaning to Mathematical Life," *New Scientist* **129**(1762) (1991), 44–45.

Schilpp, P. A. editor, *The Philosophy of Bertrand Russell*, 3rd ed., New York: Tudor, 1951.

Schuster, H., *Deterministic Chaos*, 2nd rev. ed., Weinheim: VCH Publishers, 1988.

Sierpiński, W., *On the Congruence of Sets and Their Equivalence by Finite Decomposition*, Lucknow: Lucknow University, 1954.

Stein, S., *Mathematics—The Man-Made Universe*, San Francisco: W.H. Freeman and Company, 1963.

Stewart, I., *The Problems of Mathematics*, Oxford: Oxford University Press, 1987.

———, "The Ultimate Jigsaw Puzzle," *New Scientist* **130**(1764) (1991), 30–33.

———, "Does Chaos Rule the Cosmos?" *Discover* **13**(11) (1992), 56–63.

———, "Paradox of the Spheres," *New Scientist* **145**(1960) (1995), 28–31.

———, *From Here to Infinity*, Oxford: Oxford University Press, 1996.

———, *Does God Play Dice*, 2nd ed., Malden: Blackwell Publishing, 2002.

Stromberg, K., "The Banach-Tarski Paradox," *American Mathematical Monthly* **86** (1979), 151–161.

Su, F., "The Banach-Tarski Paradox," Ph.D. Minor Thesis, Harvard University, 1990.

Sutton, C., "World of Quarks," *New Scientist* **139**(1881) (1993), 1–4 (inserted between 42 and 43).

Svozil, K., *Randomness and Undecidability in Physics*, Singapore: World Scientific, 1993.

———, "Set Theory and Physics," *Foundations of Physics*, **25**(11) (1995), 1541–1560.

———, "'Linear' Chaos Via Paradoxical Set Decompositions," *Chaos, Solitons and Fractals* 7(5) (1996), 785–793.

Twain, M., *Following the Equator: A Journey Around the World*, Hartford: The American Publishing Co., 1897.

Vilenkin, N., *Stories About Sets*, New York: Academic Press, Inc., 1968.

Wagner, C.H., "Simpson's Paradox in Real Life," *The American Statistician* **36** (1982), 46–48.

Wagner, J., *The Search for Signs of Intelligent Life in the Universe*, New York: Harper and Row, Publishers, 1986.

Wagon, S., *The Banach-Tarski Paradox*, New York: Cambridge University Press, 1985.

Wagon, S., "A Hyperbolic Interpretation of the Banach-Tarski Paradox," *The Mathematica Journal* 3 (1993), 58–61.

Wagon, S., "Preface to the Paperback Edition." In *The Banach-Tarski Paradox*, paperback ed., New York: Cambridge University Press, 1993.

Wilder, R., *Introduction to the Foundations of Mathematics*, New York: John Wiley and Sons, Inc., 1952.

Yandell, B., *The Honors Class*, Natick, MA: A K Peters, 2002.

Youngs, D., "Lost in Space," *AIMS Education Foundation* (http://www.aimsedu .org/Puzzle/LostInSpace/space.html), March 22, 2002.

Zippin, L, *Uses of Infinity*, New York: Random House, Inc., 1962.

Index

Printed in the United States
by Baker & Taylor Publisher Services